IRON AND STEEL IN NINETEENTH-CENTURY AMERICA

AMERICA

An Economic Inquiry

M.I.T. MONOGRAPHS IN ECONOMICS

IRON AND STEEL IN NINETEENTH-CENTURY AMERICA

An Economic Inquiry

PETER TEMIN

THE M.I.T. PRESS
Massachusetts Institute of Technology
Cambridge, Massachusetts

Library of Congress Catalog Card Number: 64-22211
Printed in the United States of America

Acknowledgments

I would like to thank the following people and organizations for their generous help in the preparation of this volume.

Alexander Gerschenkron helped give direction to my research in his role of advisor for the (M.I.T.) doctoral dissertation which appears here in slightly altered form. Richard Eckaus, Charles Kindleberger, Wassily Leontief, and Robert Solow read and criticized an earlier draft of the thesis. Elting E. Morison allowed me to use the results of his unpublished research on the introduction of the Bessemer process. Robert Fogel shared with me his interesting work on the ante bellum iron industry. Franklin Fisher helped to resolve some statistical problems. And Paul David, Albert Fishlow, William Whitney, and Robert Zevin offered helpful opinions, references, comments, and arguments in the course of conversations at the Economic History Workshop.

The Workshop, under the direction of Alexander Gerschenkron, provided a congenial atmosphere in which to work, as well as support for part of the time necessary for research. The Society of Fellows of Harvard University supported me for the rest of the time and placed me in a setting conducive to good work.

The *Journal of Economic History* has allowed me to reprint parts of Appendix C.

All responsibility for errors and omissions, of course, remains mine.

PETER TEMIN

August, 1964

Contents

List of Tables

Introduction

The growth of an industrial economy remains a mysterious process to those engaged in studying and describing it. It involves changes in thought patterns as well as changes in the means of production. It requires a transformation of agriculture as well as an expansion of industry. It generates — and sometimes cures — a host of problems only dimly apparent in a preindustrial world. But a complete view of this process is beyond our grasp at the moment, and the investigator must proceed by isolating specific developments, examining them in detail, and then relating them to changes elsewhere in the economy or polity.

The inquiry whose results are printed here concerns a restricted part of the United States economy, and employs a particular methodology to illuminate one aspect of the growth in this area. It is an industry study because it was felt that the growth of an industry could be analyzed relatively simply; it is a study of the iron and steel industry both because the production of iron and steel was an important part of American economic growth in the nineteenth century, and because the industry contained within itself sufficient diversity and change to provide an interesting sequence of problems. The methods used are those of economic analysis. They take as their primary data the movements of prices and quantities of goods sold, and attempt to explain these data on the basis of changes in such things as the technical requirements of production, the organization of production, and the character of demand.

It is still uncertain when the process of industrialization may best be considered to have begun in the United States, but 1830 forms a good starting point for the iron and steel industry. This industry, which was almost exclusively an iron industry at that

1

time, had yet to begin the transformation of its productive methods, raw material base, and product structure that typified the years after 1830, but the knowledge that would permit it to do so—as well as the competing iron which provided pressure for these changes—was being imported from England at that time. Ironmasters were beginning to experiment with the new methods of production around 1830; by the 1840's, the new methods were widely known, and the process of change was under way.

Similarly, 1900 forms a convenient terminating point for the story as told here. At that time the organization of the iron and steel industry was changed by the merger wave that resulted in the formation of the United States Steel Corporation. At that time also, the two temporary influences that had directed much of the development of the nineteenth century—the production of rails and the use of the Bessemer converter—had ceased to determine the direction of change of the iron and steel industry. This history is the story of the interaction of these two influences with the more gradual and cumulative processes that typify much of the industrialization of the nineteenth century. It ends as their influence waned.

The more permanent and cumulative of the forces at work upon the iron and steel industry after 1830 may be separated, following the line of argument used extensively in the text, into those deriving from the demand side of Marshall's scissors and those originating on the supply side. Forces impinging on the iron and steel industry from the demand side may be expected to be characteristic of the American economy as a whole, while forces that operate through their influence on the supply curves of iron and steel may be expected to be more characteristic of the industry itself.

The most pervasive single force acting upon the supply of iron and steel was an increasing sophistication in the use of heat. Heat is a form of energy, and its usefulness for the production of iron and steel was appreciated in the course of the nineteenth century. The iron and steel industry could use mechanical energy in much the same way as other industries, but unlike many other industries, it also needed energy for chemical reactions. This energy could be supplied in several ways, but heat was the most efficient form. The reverberatory stove, the hot blast, and the Siemens regenerative gas stove were only the most important of the means designed to capture and exploit heat for this purpose. The use of heat to induce chemical reactions led to an awareness of the heat being generated and largely wasted around the ironworks of the early nineteenth century and to arrangements to save and utilize this heat. This led to the integration of operations to facilitate the transfer of heat from one stage to the next, to the speeding of operations to take advantage of the faster reactions induced by heat, and to the widespread use of mechanical energy produced by the heat not needed for chemical transformations.

An equally strong influence was exerted upon the iron and steel industry by the growth of the American economy, which produced an ever-growing demand for iron and steel as well as a continuing demand for better quality iron and steel in more complex and heavier shapes. The increase in the demand for iron and steel that derived from the fabled growth of the American economy was almost as great as the increase in supply that derived from developments in the industry. The production of pig iron in 1900 was some seventy-five times as large as the production in 1830; even though the price was lower, this growth argues for an expansion of demand of considerable magnitude.

The growth of the American economy, however, had an influence on the production of iron and steel that may profitably be distinguished from the effects of its larger magnitude. This influence derived from the rate of growth of the economy, an analogue to the "accelerator" process in which the level of investment is directed by the rate of change of output, rather than by the level of output. An expanding economy needed transportation facilities, which meant railroads for at least the last half of the nineteenth century. The rate of building railroads was partly a function of the rate of expansion of the economy, and the rate of consumption of iron and steel for rails was a function of the rate of railroad construction. (It was also a function of the number of miles of railroads in use because of the need for replacements, which is why the demand for rails fell, but did not cease, at the end of the nineteenth century.) The size of the United States and the extent to which transportation in the late nineteenth century was dependent upon the use of the iron horse assured that one of the major products of iron and steel would be rails. In fact, over one-third of the rolled iron and steel products made in America in the quarter-century following 1860 were rails.

The production of rails had many characteristics that made it an important influence on the production of rolled iron in the years just before the Civil War, but its most important influence came through the demand for a better quality rail. The result of this demand was the introduction of Bessemer steel in the years just after the Civil War, a substance tailored for the production of rails. The Bessemer process, by which this type of steel was made, represented an unorthodox leap in the progression toward higher heat in the production of iron; this jump forms a point of discontinuity in the otherwise more gradual development of sophistication in production techniques.

The Bessemer process represented a discontinuity in many of the characteristics of iron manufacture, and the transition from iron to steel in the late nineteenth century was a complex one. The minimum economic size of a Bessemer plant was far larger than anything known before, a development which led to a different industrial structure in the production of steel than had prevailed in the making of iron. The chemical requirements were more precise than those of previous processes, creating a demand for knowledge which led to developments in other branches of the industry. And the advantages of integration were increased by these and other factors, leading to the integrated steel firm that typified the late nineteenth century.

But the Bessemer process, for all its advantages in the production of rails, was not as well suited to the production of other products or to the use of available resources as the less spectacular open-hearth process which replaced it. This process was a more logical result of the trend toward the use of higher heat that was mentioned above, and its greater flexibility may be attributed to this. The importance of rails for the iron and steel industry began to fall in the 1880's: the importance of the Bessemer process began to decline in the following decade. In the years around 1880, almost nine-tenths of the steel made was Bessemer steel, but as four-fifths of this was used to make rails, the condition could not last. By the time of the First World War, only one-third of the steel made in the United States was Bessemer steel.

The influence of rails on the production of iron and steel, then, was only temporary. The importance of rail production was greatest during the period of great geographical expansion, and the importance of the technical innovations it sponsored was also restricted in time. It is hard to say that the twentieth-century steel industry would have been unchanged if the Bessemer process had never been discovered, but the short life of the Bessemer process suggests that it might not have been too different. Nevertheless, the nineteenth-century steel industry would have been very different, and the effects of this are too complex to be known.

The story of the American iron and steel industry from 1830 to 1900 is thus the result of a combination of forces. Two gradual and cumulative trends — increasing sophistication in the use of heat and a growing demand for iron and steel — had superimposed upon them forces stemming from two more sudden and temporary developments. The growth of the economy produced a demand for rails which led to the exploitation of the Bessemer process. The Bessemer process in turn advanced the trend toward greater exploitation of the available heat and power, which facilitated the further growth of the economy and of the demand for iron and steel. It is apparent that most gradual trends are aided, obstructed, or at least altered by the presence of temporary developments that produce their own highly specialized effects. The differences between times and places may be the result of the extent to which these temporary influences correspond with the requirements of the general trends. The nineteenth-century American iron and steel industry was helped by the extraordinary importance of railroad construction; the twentieth-century industry was probably influenced as much by the demand for flat products for automobile construction.

The development of the iron and steel industry before 1900 is approached in the following pages through the medium of three main variables: (1) the quantity and composition of the production of the industry, (2) the methods of production, including the type of machinery and inputs used, and (3) the nature of the firms comprising the industry: their size, number, degree of integration, and forms of combination. Changes in these rather complex variables are analyzed by means of the time-honored distinction between supply and demand.

The separation of changes in demand from changes in supply, however, is a complex process, involving what has become known as "the identification problem." This is a problem which is concerned with the separation of the manifold interactions of different variables into distinct and well-behaved economic relations, the most obvious and important of which are the supply and demand schedules that are basic to much of economics and to this particular history. A supply schedule (or curve) is a relationship between the price and quantity of a particular commodity which tells how much of this good its producers are willing to provide for sale at any given price. The demand schedule (or curve) for this good is another relationship between the same price and quantity which shows how much of the good people are willing to buy at a given price. As these are both relations between the same variables, it is not easy to tell from a set of observations of associated prices and quantities which of them is being observed.

Additional information is necessary to identify which relation is being observed. The most common form of this information involves the way in which the schedules shift over time; the variables observed are then price, quantity, and some other variable (or variables) showing the movements of the curve. This approach to the problem is unfortunately not open to the historian of the nineteenth century as the extant data are not sufficiently complete to permit work along these lines. It is necessary to deal with the identification problem by means of assumptions about the particular relations being considered. In other words, each time the problem arises, and it arises frequently, a set of assumptions will be introduced to resolve it. The particular assumptions used will be explained in the course of the text.

Chapter 1, together with Appendix A, describes the volume and composition of the iron and steel industry's output between 1830 and 1865 as well as can be determined from the fragmentary data. The industry may be divided into two branches, one producing pig iron from iron ore and the other making wrought iron and steel from pig iron. In this early period, the first branch was expanding relative to the second, which made wrought iron almost exclusively. By the use of assumptions made to resolve the identification problem, it is asserted that this shift was due to movements of demand. The demand for cast iron was rising faster than the demand for wrought iron, and this demand was rising swiftly enough to counterbalance the faster fall of the supply curve for wrought iron than for cast iron and to lead the industry toward the production of cast iron. (In Marshallian supply and demand curves, price is measured along the vertical axis and quantity along the horizontal. Consequently, an increase in the quantity people will *demand* at a given price appears as a *rise* in the demand curve — due to its negative slope — while an increase in the quantity people will *supply* at a given price appears as a *fall* in the supply curve — due to its positive slope.)

The opposing movements of demand and supply curves during the growth of "the new iron industry" are discussed in Chapters 2 and 3. Due to the paucity of quantitative data, the analysis of demand shifts in Chapter 2 is taken almost entirely from literary and qualitative sources. As such, it can give no more than a general idea of the trends in demand in the ante bellum years; its main purpose is to place the rising demand for rails after 1850 in perspective. Chapter 3 addresses itself to the problem of why the supply curve for wrought iron was falling more rapidly than that for cast iron. More exactly, Chapter 3 attempts to show why the technological advances hinging on the use of mineral fuel in the blast furnace were largely ignored before the Civil War west of the Allegheny Mountains where half the iron industry was concentrated. Since the wrought-iron sector of the industry was adopting new methods at a rapid pace in the 1840's and 1850's, this demonstration explains the relative movements of the two supply curves. It is asserted that the use of mineral fuel in the West was not profitable and that this was remedied about the time of the Civil War by an improvement in the quality of pig iron made from coke, deriving either from the discovery of better coal deposits or a growing awareness of the importance of good coal. The less stringent requirements for fuel to make wrought iron had permitted the iron industry to advance in this area well before the Civil War.

The first three chapters, then, describe the movements of the first two variables listed above. The remaining two chapters of Part I complete the analysis of the methods of production, and show their relationship to the nature of ironmaking firms. Chapter 4 deals with the manufacture of pig iron in the blast furnace, showing the effects of the change to mineral fuel on the size and character of blast-furnace establishments. Chapter 5 considers the changes in the production of wrought iron induced by the rise of rolling mills. The technological advances adopted in these years are described, and their interaction with the nature of firm and industry is analyzed. In both parts of the industry, the changes denoted here under the rise of "the new iron industry" led to an increase in the scale of iron-producing firms together with a decrease in the vertical integration of the industry. A new trend toward integration had started, however, by the end of the period, originating from developments in the wrought-iron branch of the industry. These developments were allied with the production of iron rails and the growth of firms centered around rail mills; they show the influence of rail-making on the iron industry before the introduction of the Bessemer process.

The narrative is broken at 1865, the time of the introduction of the Bessemer process. Part I treats the developments that were contemporary with the development of rail production; Part II shows those developments that followed the introduction of new methods for making rails.

Chapter 6 opens the second part with a description of the new methods of making steel. The first of these was the Bessemer process, with its spectacular nature and its problems of high heat and complicated machinery. The innovations of Alexander Holley helped to resolve these problems and make the Bessemer process profitable in America. Soon after the Bessemer process came the open-hearth process, although it had higher costs than the Bessemer process in the 1870's. In the following decade, basic furnace linings that could be used in conjunction with both the Bessemer and open-hearth process were introduced, and the basic open-hearth process proved itself to be the most flexible of the processes available by the turn of the century, adaptable to both the resources and the demand present in America.

But the Bessemer process was the first, and for a while the best, process invented for making steel directly from pig iron. Its use expanded in the 1870's, and its technical characteristics shaped the steel industry. The minimum economic size of a Bessemer plant was sufficiently larger than the size of an efficient iron rolling mill to force a new kind of firm and industry structure upon the steel industry. All the early Bessemer firms were integrated, and they were large enough to think in terms of organizing the market for steel in order to gain monopoly profits. The integrated firms exploited the advantages of high heat and of integrated operations; the result was a growth in efficiency as well as a growth in the size of the minimum cost plant. These firms tried time and time again to capitalize on the small number of firms in the industry by collusion; but only with rails were they successful, and even with rails their effectiveness was limited. Chapter 7 is concerned with the changes within the firms themselves; Chapter 8 is concerned with their interaction.

The steel industry brought to life by Bessemer's discovery was superimposed upon the already existing iron industry, which was gradually superseded by it. The last two chapters of Part II deal with the iron and steel industry as a whole, showing the effects of the growing steel production on the composition of the industry. Chapter 9 surveys the two principal raw material inputs of the industry: ore and coal. The effects of changing deposits of these materials are integrated with the effects of changing demand. The discussion of Chapter 3 finds its sequel in the investigation of the relative costs of coke and anthracite smelting in 1890. It is shown — Appendix B contains the actual statistical proof — that by 1890 coke was a definitely superior blast-furnace fuel, due to its adaptability to the innovations introduced by the integrated steel firms. Chapter 10 deals with the outputs of the iron and steel industry. It describes the changing allocation of pig iron production, and the changing nature of the products made from wrought iron and steel. The material presented earlier on the supply curves for steel is used to resolve the identification problem and to show the effects of shifting demand on the industry's output. The supply curve for steel was falling at the same time as the importance of rails was waning. The growing importance of steel and the (relatively) declining demand for rails together helped to shape the composition of production at the end of the nineteenth century.

The second part of this history, then, closes with a consideration of the iron and steel industry's production, in much the same fashion as the first part opens. But while the data for the ante bellum years permit only a resolution of output between the two branches of the industry, the data for the end of the century allow a breakdown by product. The organization is thus not strictly circular; it opens and closes with the consideration of the more general influences that determined the production of the industry, but the treatment differs, even as the forces that were operative in 1830 and 1900 differed from each other.

Although the parts of this work are divided chronologically at 1865, the chapters within the parts — as can be seen from the foregoing discussion — are arranged topically, each chapter within a part tracing essentially the same time period from a different point of view. Appendices A and B supplement Chapters 1 and 9; Appendix C contains quantitative data of general interest that is used throughout the text.

PART I

THE NEW IRON INDUSTRY,
1830–1865

1

The Start of a New Era

Before The New Era

The first attempt by Europeans to produce iron in the United States was initiated in 1619. Unfortunately, this enterprise was destroyed by Indians in 1622, probably before it made any iron. A more successful attempt took place in the Massachusetts Bay Colony in 1645, and it is from this event that we may date the American iron industry.[1]

By 1700 the American colonies were producing about 1,500 tons of iron annually, or something under 2 per cent of the probable world production and perhaps about 10 per cent of the probable British production. The British industry was based on imported iron from Sweden and elsewhere, even though imports of iron were taxed heavily. The first regulation to differentiate between American iron and other imports was an act of 1750, which (as amended in 1757) provided that American iron could be admitted to England free of duty. As this act was an attempt to increase the supplies of raw material for the English industry without destroying the colonial market for its products, it carried with it the restriction that no new ironworks producing finished products could be built in the colonies after June 24, 1750. The prohibited classes of ironworks included slitting mills, which made nails, plating mills, which made hammered sheets and tin-plate iron, and steel furnaces, which made blister steel.

Despite several changes in the details of the laws about the colonial iron industry, their intent remained faithful to the act of 1750. Nevertheless, the British regulatory policy was

[1] James M. Swank, *History of the Manufacture of Iron in All Ages* (second edition; Philadelphia: American Iron and Steel Association, 1892), pp. 103-112.

only moderately successful. While imports from the colonies rose, they never attained a level of high significance. At their peak in 1771, American exports of bar and pig iron were about 2,000 and 5,000 tons, respectively, compared with British imports of 43,000 tons of bar iron from other sources (but no other imports of the far less valuable pig iron). In addition, American production of iron rose from about 10,000 to about 30,000 tons in the same period. This iron no doubt found its way into consumption within the colonies by way of ironworks prohibited by colonial law. The inability of the English legislators to determine the direction of the colonial industry can be attributed to their internal conflicts of interest, but it also derived from their inadequate knowledge of the colonies and lack of appreciation for the rapidly growing colonial demand. In any case, the production of iron in America on the eve of the Revolution may have been as high as 15 per cent of world production, a proportion that was not again equaled until well into the nineteenth century. And while this high production was largely for domestic consumption, the colonies were net exporters of unfinished iron.[2]

In the last quarter of the eighteenth century the British iron industry went through a technological revolution that abruptly altered its relation to the world and to the American iron industry. This revolution started with the introduction of coke as a blast-furnace fuel, replacing the time-honored fuel of charcoal. The use of coke may have been proven a technological possibility as early as 1709 by Abraham Darby, but its general commercial feasibility had to wait upon further developments of technique, taking another half-century. Once the skill of making iron with coke had been advanced to the proper level, however, the transformation of the British iron industry was rapid. In 1788 only 24 out of 77 pig iron furnaces in Britain used charcoal; in 1806, only 11 out of 173.[3] The new technology not only removed England from dependence on its depleted forests and enabled it to exploit its abundant coal, it also opened the way for further improvements in making iron. Britain changed from

[2] The preceding discussion has been adapted from the exposition in Arthur Cecil Bining, *British Regulation of the Colonial Iron Industry* (Philadelphia: University of Pennsylvania Press, 1933).

[3] H. R. Schubert, *History of the British Iron and Steel Industry from c. 450 B.C. to A.D. 1775* (London: Routledge and Kegan Paul, 1957), p. 335; Thomas S. Ashton, *Iron and Steel in the Industrial Revolution* (Manchester: At the University Press, 1924), p. 99; John Gloag and Derek Bridgwater, *A History of Cast Iron in Architecture* (London: George Allen and Unwin, 1948), p. 42.

a high-cost producer to a low-cost producer and, as a result, from a net importer to a net exporter of iron.

The United States, on the other hand, continued to employ the old technology. It became a high-cost producer and, despite the imposition of tariffs, a net importer of iron. Nevertheless, the iron industry in this country grew until in 1830 the production of pig iron was close to 200,000 tons. This figure, the product of a free trade convention in Philadelphia and a protectionist convention in New York, represents actual data for about 45 per cent of the furnaces "reported" and imputed production for the rest. But as there was no census of manufactures in 1830, it is the best we have.[4]

It would be unwise to make a strict statement about the rate of growth in the years before and immediately after 1830. The extant data are too fragmentary and come at too infrequent dates for a cited growth rate to have much meaning. But it may not be too misleading to suggest that before the 1840's, for which more exact data have survived, the iron industry of the United States was expanding only slightly more rapidly than the population. If the census figure of 54,000 tons of pig iron produced in 1810 can be believed, the per capita production of iron in the United States did not rise between the Revolution and that date.[5] In any case, the iron industry, like the economy, was only making its first tentative steps toward expansion. It was still using the old techniques and the traditional fuel, charcoal, inherited from previous centuries, much as the economy was largely concentrated in its familiar location. The new techniques that were being brought over from England had all the hazards of a new frontier, and as Leland Jenks says of the frontier in 1830, "It was a speculation, not a going concern."[6]

Before considering the "speculation" in new ironmaking techniques that occupied the American iron industry after 1830, we must ask how the industry arrived in the situation in which it found itself at that time. The American iron industry in 1830 was still operating almost exclusively on the basis of a traditional technology, despite the very successful exploitation of a newer technology in England, an exploitation that threatened to ruin the American industry if unrestricted trade were permitted.

[4] For a complete discussion of these figures, see Appendix A.

[5] Appendix C, Table C.1; Appendix A.

[6] Leland Hamilton Jenks, *The Migration of British Capital to 1875* (New York: Alfred A. Knopf, 1927), p. 73.

Why was the American industry trying to protect itself against the foreign methods of making iron in preference to importing them?[7] There is not sufficient information on costs to answer this question fully, but an examination of the technology in question provides a plausible explanation.

Most of the iron used at this time was in the form of wrought iron, often called bar iron after the shape in which it was most commonly sold. Wrought iron could be forged into different shapes due to the relatively pure state of the iron of which it was composed. It was made, therefore, by removing all elements but iron from iron ore. The primary offending element was oxygen, and it had been discovered long before the nineteenth century that wrought iron could be made most profitably in a two-stage process. The direct process of making wrought iron from ore did not die out until well into the nineteenth century, but the products of "bloomary forges" which made iron in this way were already only of minor importance in 1830.[8]

The first step in the indirect manufacture of wrought iron utilized a blast furnace, a container in which iron ore and fuel were burned. The oxygen was banished from the iron ore, but its place was taken by part of the carbon from the fuel. The resultant alloy of iron and carbon was not malleable, although it had a relatively low melting point and could be cast into different shapes. This alloy was known as pig iron when it was used as an intermediate product and as cast iron when it was used in iron castings.

The second step in this indirect process was to expel the carbon from pig iron to make wrought iron. This process can be accomplished by the use of great heat, which causes the desired chemical reactions to take place, but the technology of the early nineteenth century did not have the capability of containing large quantities of metal at this heat or the knowledge of

[7] This is the opposite question to that considered by H. J. Habakkuk in *American and British Technology in the Nineteenth Century* (Cambridge, England: At the University Press, 1962). Habakkuk tries to explain why American technology in manufacturing industries was ahead of British technology in the first half of the nineteenth century. The iron industry, however, was not ahead of Britain in technological development at this time, and the discussion here is based on that fact. The discrepancy between these two treatments appears to have arisen because Habakkuk took his cue from the light manufacturing industries of New England which differed in their technological development from the heavier industries of the Middle Atlantic states.

[8] See the discussion of wrought-iron production later in this chapter.

its effects. Mechanical labor can be substituted for heat in the refining of pig iron into wrought iron, and the method of refining that the nineteenth century inherited depended upon this substitute. The iron alternately was placed in a refinery fire and beaten with a hammer on an anvil. The forges in which this process was carried out could not process large quantities of iron, and the process required a large quantity of labor to compensate for the small amount of heat that the iron received in its frequent trips to and from the fire.

In the course of the eighteenth century, inventors in England changed both steps in the indirect process of making wrought iron.[9] The traditional fuel for the blast furnace — charcoal — was replaced by coke, and the traditional method of refining pig iron by hammering was replaced by a method known as puddling. These innovations helped the English iron industry; the question before us is why they were neglected by American ironmasters. The question of fuel for the blast furnace will be postponed until the detailed discussion of Chapter 3. The point of interest here is that the form in which the new fuel was introduced in England was not suitable for use in America; auxiliary innovations were needed, of which the most notable was the hot blast. Differences between the known resources in the two countries resulted in differences in relative prices, and auxiliary innovations were needed to make the English innovations profitable at American prices.

The process of puddling involved the use of a reverberatory furnace. In this type of furnace, the iron was separated from the fuel by a low wall, and the flames from the burning fuel were conducted over the iron. The lack of contact between the fuel and iron meant that a more impure fuel than the traditional charcoal could be used without contaminating the metal. This fuel was coke: too impure for use in refining by hammering, but not for use in puddling.[10] And the separation of the iron and the fire enabled the iron to be manipulated without removing it from the heat. The puddler operated upon the iron through ports in the wall of the furnace. His labor was commuted from lifting and hammering to a sort of stirring activity: puddling.

The increase in heat obtained by use of a reverberatory fur-

[9] See Ashton for a detailed description of the innovations and their invention.
[10] Louis C. Hunter, "The Influence of the Market upon Technique in the Iron Industry in Western Pennsylvania Up to 1860," *Journal of Economic and Business History*, 1 (February, 1929), 241-281.

nace was enough to dispense with the labor of hammering, and it was discovered that the use of hammers to shape the resultant wrought iron could be dispensed with also. The replacement for hammers was grooved rolls, which could be used to make many shapes of uniform cross section. The use of grooved rolls was closely associated with that of puddling, and the two innovations were usually adopted together in the United States.[11]

But although the increase in heat obtained by leaving the iron in the furnace continuously was enough to reduce the labor required to refine it, it was not enough to keep the iron in a molten state. The melting point of wrought iron is higher than that of cast iron, and as the transformation to the former took place, the iron changed from a liquid into a pasty substance. Because puddled iron could not be kept molten, it could not be rid of the bits and pieces of nonmetallic slag that it picked up in the course of puddling. Much of the slag was expelled by "squeezing" the iron after it was removed from the furnace, but a nonhomogeneous composition remained a hallmark of wrought iron.

Puddling and rolling saved labor and fuel costs. The saving was enough to induce their general adoption in England, but not in America where costs were different. The problem was that puddling had a few offsetting defects that had to be balanced against its cost-saving attributes. One of these difficulties was the large wastage in the original process, equal to about one-half of the iron put into the furnace. This was partly due to the use of a sand bottom in the puddling furnace, following the practice of casting iron in sand. In 1818, the practice of using an iron bottom in the puddling furnace was introduced, which greatly reduced the amount of waste in puddling (to about seven hundredweights per ton of bar iron produced), increased the output of the furnace, and even improved the quality of the resultant iron.[12] This innovation permitted the puddling process to be used to advantage in America, both by reducing costs of puddling and by increasing its ability to use

[11] This refers to the general adoption of puddling and rolling, not their initial introduction. See Chapter 5.

[12] Harry Scrivenor, *History of the Iron Trade* (London, 1854), p. 252; J. C. Carr and W. Taplin, *History of the British Steel Industry* (Cambridge: Harvard University Press, 1962), p. 98n. The use of iron replaced an "acid" lining by a neutral one; this permitted phosphorus, a troublesome trace element, to be drawn off in the slag. See the discussion of linings in Chapter 7.

inferior irons. It took a few years for news of the innovation to spread to America, but by 1830 the stage was set for an improvement in the competitive position of the American iron industry.

Pig Iron Production

It is to this story that we now turn. As the story is a complex one — it will occupy the rest of this history — we begin slowly by asking what was produced by the American iron industry between 1830 and 1865. This production may be compared with the imports of British iron to give an idea of the position of the American industry and of the pattern of consumption of the American economy. To inquire further into this pattern, we look at the composition of the iron industry's production as well as at its volume, although limitations of the data force us to consider only the main outlines of this composition: the division between cast and wrought iron. This chapter closes by separating the effects of shifting supply curves from shifting demand curves on the direction of growth of the iron industry, setting the stage for the more detailed considerations of supply and demand to follow.

But before we start, it is necessary to say a word about the role of steel in the ante bellum period. The previous discussion has been couched exclusively in terms of iron because steel was an expensive, minor product at this time. It was made by heating wrought iron to replace some of the carbon removed by the transformation from cast to wrought iron. The product of this process was "blister" steel (so called because of its appearance) which shared with wrought iron the nonhomogeneous composition that was a reflection of their unmelted history. It was possible to melt steel by the start of the nineteenth century, and small quantities of "crucible" steel were made by melting blister steel in small clay crucibles, each one containing about sixty pounds. This was the sole exception to the inability to contain melted steel at this time, and its importance was not great. The product was too expensive to be widely used, and suitable clay was too hard to find for the process to be widely disseminated.[13] Only with the invention of the Bessemer

---

[13] See Ashton, pp. 54-59, for a discussion of the introduction of crucible steel-making in eighteenth-century England.

process after the middle of the nineteenth century and its improvement by Holley in 1870 was the problem of containing molten steel and thereby cheapening its manufacture satisfactorily solved.

We do not know much about the volume of pig iron production in the 1830's except that it rose. It rose again in the 1840's, and information for this decade has survived. The production of pig iron in the United States doubled between 1840 and 1847, although it probably fell between 1840 and 1842, a doubling that was accomplished by means of an extensive boom in blast-furnace construction.[14] The interesting thing about the boom in furnace construction was that it was a boom in the construction of the traditional charcoal-using furnaces as well as the newer mineral coal furnaces. The principal iron-producing region of the United States, which centered on Pennsylvania in the nineteenth century, was divided into two regions by the Allegheny Mountains, regions which were known as "the East" and "the West" to ironmasters, irrespective of the movements of the economy as a whole. Expansion in the East was accompanied by an increasing exploitation of the coal of that region for the purpose of making pig iron; expansion in the West was accomplished almost exclusively by means of the traditional charcoal-using technology. Because the new ironmaking technology was ignored by a major part of the industry, three-quarters of the pig iron made in 1847 was still made with charcoal.[15] The boom of the 1840's did not lead to a technological transformation of American pig iron production.

This boom was succeeded by a depression in the American iron industry that extended into the early 1850's. The depression was succeeded in turn by a mixed prosperity in which the production of pig iron rose, but did not greatly exceed its previous peak. As noted in Appendix A, the result of the 1850's was the replacement of about half the charcoal pig iron production with the production of pig iron with mineral fuel, not the extension of production as a whole.

[14] Appendix A contains a discussion of the production data; the production in 1847 was in the neighborhood of 700,000 to 750,000 gross tons. The data on furnace construction come from the tabulation of Pennsylvania blast furnaces with dates of construction (and much other information) compiled by an 1849 convention. See Convention of Iron Masters, *Documents Relating to the Manufacture of Iron, Published on Behalf of the Convention of Iron Masters which Met in Philadelphia on the 20th of December, 1849* (Philadelphia, 1850), Tables.
[15] Appendix A.

The cyclical fortunes of the American iron industry in the 1840's and 1850's could have been the result either of movements of supply or of demand curves. But there was no reason for the supply curve for pig iron to shift precipitously in this period; there were no shortages of raw materials, no natural disasters, no successful business combinations. The general movement of the supply curve was downward, and the continuing use of traditional methods of production shows that the movement was not swift. The variation in the condition of the iron industry in this period therefore must be attributed to changes in the demand for pig iron.

The depression of the early 1840's was part of the general depression in American business conditions that followed the boom years of the previous decade. The depression of the years around 1850 was more specific to the iron industry. The main source of distress for the American iron industry appears to have been the low prices at which British iron was being imported, a development that was caused primarily by the end of the English railway boom and not by conditions in America.[16]

In considering the mid-1850's, the demand for iron may conveniently be split into two parts: the demand for iron for rails and the demand for other iron. The former part of demand rose in the 1850's, but was satisfied largely by British producers; the latter part rose rapidly in the 1840's, but not in the 1850's. The production of iron rails in the United States rose steadily in the 1850's, as American iron producers tried to fill the growing demand, but it did not reach any great volume. The major part of the demand for rails during the railway boom of the early 1850's was filled by British rail makers who supplied over three-quarters of the iron rails consumed by American railroads in these years.[17]

The dominance of English suppliers in the railway boom had two causes. The British in this era were the low-cost suppliers of the industry, particularly of bar iron, as they had been for many years. They also catered to the American market by producing a specially cheap product for export, which contrasted sharply with the very expensive but better quality products made in America by the older, more painstaking methods. This was profitable because the American railroad

[16] See J. H. Clapham, *An Economic History of Britain*, Vol. I (second edition; Cambridge, England: At the University Press, 1930), pp. 427-428.

[17] Appendix C, Tables C.6 and C.13.

builders were interested in initial costs and did not worry about the costs of maintenance. Consequently, they imported English rails even though they disgusted American ironmasters. As Abram Hewitt, the key member of Cooper and Hewitt which owned and managed one of the largest ironworks in the United States, the Trenton Iron Company, noted in 1867, "The vilest trash which could be dignified by the name of iron went universally [in Welsh ironworks] by the name of the American rail."[18] The other part of the explanation is that the British would not only supply cheap rails, they would supply them on credit. Capital could be obtained more easily in England than in the United States, and the American railroads of the 1850's were built of British iron at least partly because they were built with British capital.[19]

As the demand for rails did not produce a large demand for pig iron in America (the British iron industry supplied its own inputs by then), the demand for pig iron in the United States came predominantly from the nonrailroad sectors of the economy. And if, despite the return of prosperity to the industry in these years, the production of iron failed to increase over its former level, this failure must be attributed to a lack of demand. For those sectors of the economy aside from railroad construction which used iron, the boom of the 1850's was not a large one. It was truncated by the crisis of 1854, and the production of iron did not rise rapidly, although it also did not waste away.

It should not be supposed that the iron industry was happy about the troubles attending these cyclical shifts in demand, which appeared to be the effects, as they were in part, of English imports. The hero who could vanquish this villain was obviously the protective tariff. The American iron industry and interested commentators were not slow to see the essentials of this melodrama, involving an already familiar hero, and were not laggard in publicizing their views. The depressions having their low points in 1842 and 1851 called forth large amounts of protectionist complaint; the growth of the middle 1840's was shown

---

[18] Abram S. Hewitt, "The Production of Iron and Steel in Its Economic and Social Relations," *Reports of the United States Commissioners to the Paris Exposition, 1867* (Washington, D. C., 1870), II, 13. See also *Hunt's, 33* (1855), 493; 35 (1856), 499. The full titles of frequently cited periodicals can be found in the Bibliography.

[19] Jenks, pp. 175, 257.

to be the result of the tariff of those years, while the difficulties
in copying that boom in the following decade were shown to
be the result of the lower tariff of that era. And if this monistic
explanation looks puerile and a bit pernicious to the disinterested
spectator of a century later, it must be admitted that to men
who were intimately involved in the problem from the point
of view of palliative policy rather than historical knowledge,
the idea of protection offered the most obvious solution and
the one most easily effected. It is doubtful whether the modern
economist could think of a better policy with the same possibility of implementation.

Protection as an idea has deep roots in the American tradition,
and the iron industry had shown an interest in tariff policy
before the difficulties of the 1850's. The two conventions of
1831, one free trade and one for protection, in which iron men
figured prominently, have already been mentioned. An 1842
estimate of iron production in Pennsylvania which formed the
basis of later extrapolations for the middle 1840's was gathered
in the depths of depression: "The intention was to shew the
effect of the prosperity of our manufacturers in creating a home
market."[20] This plea for protection was a popular one; its
strongest supporter in the iron industry was Henry Carey who
employed it in his writings and made it the basis of much of
his work.[21] The argument was that imports deprive American
workingmen of their jobs and hence of their ability to purchase.
Lack of demand then depresses other industries and all suffer
because of free trade. This obviously ignored the fact that
imports must be paid for, but is correct in the short run if the
importing country employs short-term capital imports to pay for
its purchases and there is no stimulus to production from the
payment for imports.

The evidence on which it was asserted that changes in the
tariff initiated this mechanism in the 1840's and 1850's is doubtful; it consists largely in the coincidence of the protective
tariff of 1842 and the recovery from the depression of that time,
and, later, the almost coincidence of the removal of the protec-

[20] C. G. Childs, *The Coal and Iron Trade, Embracing Statistics of Pennsylvania, a Series of Articles Published in the Philadelphia Commercial List, in
1847* (Philadelphia, 1847), p. 18. [Reprinted in *Hunt's*, 16 (1847), 586-593.]

[21] Frank W. Taussig, *The Tariff History of The United States* (fourth edition;
New York, 1898), pp. 109-122.

tive tariff in 1846 and the fall in iron prices due to imports in 1849. While a rough correlation between changes in the tariff and changes in business conditions existed, a correlation does not indicate causation. And it appears likely that the causation worked in a direction opposite to the one assumed by Henry Carey. When conditions were bad, the producers of iron held conventions, sent memorials to Congress, and tried to raise the tariff. As prices were low and the industry was demonstrably in trouble, Congress might grant the tariff increase. However, when the condition of the economy and the industry improved in the normal course of events, the users of iron complained in their turn to Congress about the high prices they were being forced to pay for iron. Congress then was more likely to hear their complaints than the opposition of the prosperous iron industry. And if the tariff was removed (and even if it was not), when hard times came again, the iron industry, seeing evidence of the pernicious effects of free trade, held more conventions.

Accordingly, the next rash of conventions of the iron industry after 1842 was in 1849, after the (iron) boom of the 1840's had ended, an event that had more to do with the end of the British railway boom than with the tariff change of 1846. It was at this time that Carey presented his famous series of production estimates illustrating the rapid growth in the protectionist years, together with other arguments for protection.[22] A convention in Philadelphia collected the data that enable us to check Carey's estimates, together with other evidence designed to show the existence of a depression. They reprinted a letter from Cooper and Hewitt to the *Journal of Commerce*, for instance, asserting that the United States could not compete with England in the production of wrought iron (due to the high cost of labor).[23]

A New York convention, called, like the one in Philadelphia, to modify the tariff, summarized the state of the industry as its members saw it: "We feel safe in asserting that of the large amount of machinery in this country for the manufacture of iron, not more than 33 per cent thereof is in operation at this time, and this proportion is daily diminishing, and a considerable share of this is kept in motion more for the sake of keeping

[22] Henry C. Carey, *The Harmony of Interests*, printed in *Miscellaneous Works* (Philadelphia, 1872).

[23] Convention of Iron Masters, *Documents*..., Tables and pp. 54-68; Appendix A.

up business organization than from any profit derived from it."[24]

This depression gave way to the mixed picture of the mid-1850's: reports of prosperity, mixed with notices that conditions were not all good and the background knowledge of the decline in the production of charcoal pig iron.[25] This period worsened after 1857, and the iron industry went into a difficult period of small fluctuations that lasted through the first years of the Civil War. Abram Hewitt, of Cooper and Hewitt, may be cited as a responsible observer: "Now [in 1861], the channels of industry appear to be dried up; the demand for iron is restricted to the uses of war; sales are almost impossible, and prices are unremunerative."[26] If conditions were not as bad as Hewitt imagined, as may be inferred from the failure of iron production to fall greatly after 1857, the period was also not one of unbounded prosperity.

Wrought-Iron Production

The varying production of pig iron in the ante bellum era was due to fluctuations in the demand for iron stemming from conditions both at home and abroad. The fortunes of the other branch of the iron industry, the branch that completed the two-step process for making wrought iron, are not known with the same confidence as those of the pig iron producers. The increasing production of rails in the 1850's suggests that the producers of wrought iron were prospering, but this is little more than a suggestion. We must content ourselves with observing and explaining the trends in wrought-iron production, specifically its trends relative to the production of pig iron. The absolute volume of wrought-iron production is not known with accuracy, and it is easier to follow and explain the course of wrought-iron production seen as a part of the total iron industry's product than as an isolated phenomenon.

According to the conventions of 1831, about 85 per cent of the pig iron produced was converted into wrought (or bar) iron. In this way about 90,000 tons of bar iron was made, which should

[24] Convention of Iron Workers, *Proceedings of a Convention of Iron Workers Held at Albany, N.Y., on the 12th day of December, 1849* (Albany, 1849), p. 25.

[25] *ARJ*, 26 (1853), 666; 27 (1854), 542; *Hunt's*, 29 (1853), 589; American Iron Association, *Bulletin* (Philadelphia, 1857-1858), p. 33.

[26] Trenton Iron Company, *Annual Report of the Secretary* (Abram S. Hewitt), 1861 (New York, 1861), p. 7.

be contrasted with the 8,000 or so tons of bar iron which the convention listed as the result of the direct reduction of ore.[27] The 8,000 tons were made because the small scale of the direct process made it preferable in some isolated areas. As transportation improved, this advantage lost its importance and the proportion of bar iron made in this way declined. It may safely be ignored in the following discussion of the allocation of iron production.[28]

Nevertheless, the approximate nature of the aggregate figures needs to be remembered. The underlying data were undoubtedly incomplete, and only an incomplete distinction was made between bloomary forges, making wrought iron from the ore, and refinery forges, making wrought iron from pig iron. There is no way of avoiding this confusion in the sources, just as there is no way to increase their incomplete samples. Wrought iron declined from its pre-eminent position in the manufacture of iron articles, but whether that took place after 1830, as shall be assumed here, or before, is not known. Similarly, the time at which the iron industry switched from its dependence on the small-scale, direct method for the production of wrought iron to the more efficient indirect process can only be estimated. These matters of timing are not important for an investigation of the events following 1830.

In 1830, then, about 85 per cent of pig iron production was destined to be converted into wrought iron. By 1850, this pro-

[27] Friends of Domestic Industry, *Report of the Committee on the Product and Manufacture of Iron and Steel of the General Convention of the Friends of Domestic Industry assembled at New York, October 26, 1831* (Baltimore, 1832), pp. 15-16.

[28] Despite the seeming definiteness of the Friends of Domestic Industry's report, there is considerable doubt about the extent of this direct production. Small in scale, scattered in location, it was known only within wide limits. McLane's report on Manufactures, for example, reports a production of bloomary iron in western Pennsylvania alone that is close to the national total of 8,000 tons reported by the New York convention. But McLane's Report also shows a somewhat higher proportion of the pig iron made being used for castings, together with a slightly higher waste factor in forge production of wrought iron, and these changes offset the effect of the increased direct production in aggregate calculations.

See U.S., Congress, House, Executive Document No. 308, 22nd Cong., 1st Sess., "Documents Relative to the Manufactures in the United States," collected by the Secretary of the Treasury, Louis McLane (1833), II, 646. The calculations were done for western Pennsylvania only, due to incompleteness and irregularity in the other data. See Chapter 4.

portion had fallen to less than half, where it stayed, roughly, until the end of the Civil War.[29]

Although this implies that the production of castings rose more rapidly than the production of pig iron, it does not imply that the production of wrought iron rose correspondingly slower. For this would only be true if the technological coefficients remained constant over time. And while it is a good assumption to say that the volume of castings can be approximated by the volume of pig iron used for this purpose, the same is not true for wrought iron. The production of castings did not involve the transformation of the material used, and the technology of iron casting did not change substantially in this period; the production of wrought iron from pig iron did involve a transformation of the material being used, and the technology of this transformation was changing due to the introduction of puddling and rolling. It was noted above that the use of iron bottoms in puddling increased the quality of the wrought iron produced. A corollary of this was that inferior iron could be used in puddling, which implied in turn that a far greater volume of scrap could be used in puddling than in the older refining process. As a result of this characteristic of puddling, its introduction altered the ratio of the volume of pig iron used to make wrought iron to the volume of wrought iron produced. This ratio fell from about 1.3 in 1830 to about 0.9 in 1850 and 0.75 in 1870. Consequently, the production of wrought iron almost kept pace with the production of pig iron. The data show a fall in the ratio of the two volumes from two-thirds to one-half between 1830 and 1850 and a rise following that date, as shown in Table 1.1.

Table 1.1 summarizes the preceding statements and, in the last row, shows an important extrapolation from them. The roughly parallel movements of the volumes of wrought iron and pig iron produced did not indicate parallel movements in the production of wrought iron and of castings. As the wrought-iron-producing sector of the industry decreased its dependence on the pig iron producers in the years before 1850, its place was taken by the makers of iron castings. The result was not only a stable ratio between the production of wrought and pig iron, as shown in the third row of the table, but—because of the change in the ratio in the first row—a rapid rise from 1830

[29] Table 1.1, first row.

TABLE 1.1
RATIOS OF VARIOUS VOLUMES, 1830-1870

	1830	1850	1870
Pig iron used to make wrought iron / Pig iron produced	0.85	0.45	0.55
Pig iron used to make wrought iron / Wrought iron produced	1.3	0.90	0.75
Wrought iron produced / Pig iron produced	0.65	0.50	0.75
Pig iron used to make castings / Wrought iron produced	0.25	1.1	0.60

Source: Friends of Domestic Industry, pp. 15-16; Appendix C, Table C.8 (rounded). All ratios are ratios of tons, not values.

to 1850 in the volume of castings made relative to the volume of wrought iron produced, as shown by the ratio in the last row.[30]

This trend toward castings reached its peak near 1850, and Table 1.1 shows that the trend of production had been reversed by 1870. The sharpness of this short-lived trend seems to indicate a major transformation in consumption patterns, but it is partially a reflection of the changing relation of the American iron industry to the American market. It was mentioned above that English imports were making serious inroads on American markets by 1850, and that one of the principal products affected was rails. Rails were made of wrought iron, and the English imports were mostly of wrought iron. As a result, the ratio of the consumption of castings to the consumption of wrought-iron products went only from 0.25 in 1830 to about 0.7 in 1850.[31]

The trend in the pattern of consumption, therefore, followed the direction and the timing of the trend in production, but because of the character of imports, it was of smaller magnitude. In fact, part of the trend in production was a compensation for the difference between the composition of imports and of consumption. The British were exporting wrought-iron products to

[30] The volume of pig iron used to make castings is, as noted in the text, a good proxy for the volume of castings made as there was little waste in the fabrication of castings.

[31] See Appendix C, Tables C.8 and C.13. It was assumed that fiscal year 1850 was the same as census year 1850. The ratio for 1870 was slightly below 0.6.

the United States at a price that the American iron producers could not match. They consequently turned to the production of cast-iron articles, returning to the production of wrought iron when the composition of consumption no longer demanded as many castings and when they were better able to compete with the British wrought-iron producers. The fall in the ratio of pig iron used to make wrought iron to wrought iron produced reflects the technological transformation that started the American iron industry toward a competitive cost position.

The causal network behind Table 1.1 went, therefore, as follows: the pattern of consumption altered in favor of cast-iron articles at the same time as increasing quantities of wrought-iron articles were being imported, forcing the American iron industry to alter the composition of its output. The last row of Table 1.1 has been treated as an extrapolation from the other ratios; in reality, it changed as a result of forces not shown in the table, and the other ratios followed from its movements (and those of the technological ratio shown in the second row). In order to understand the changing allocation of iron production in America, we must inquire into the nature of these forces.

Supply and Demand

The forces that influence the composition of consumption can be separated into those that operate by shifting the supply curve for some or all of the products in question and those that alter the demand curves for these products. This separation, while easy to make in theory, is hard to make in an actual situation due to the "identification problem." Data on prices and quantities will not solve the problem as they do not provide enough information to identify movements in one or the other class of curves. Additional information is necessary, which cannot be introduced here in the form of statistical series. Instead, we use qualitative information to support an assumption that permits the resolution of the price and quantity data into movements of supply and demand curves.

The data on prices and quantities, however, must be reviewed first. The data on the relative quantities of wrought and cast iron consumed have been presented above; there was a pronounced rise in the relative quantity of cast iron consumed before 1850 and a fall after then. The extant data on prices show a unidirectional trend with the price of wrought iron falling relative to that of cast iron. The price of pig iron had little trend in the period before the Civil War disrupted prices, while the price of bar iron fell. This was placed against a background of a trendless wholesale price index for all commodities which nevertheless shows the well-known transitory rises in the 1830's and 1850's and trough in the 1840's.[32] The trend in relative prices continued through the price fluctuations of the Civil War.

In the West, the pig iron price maintained a level of about $30 a ton in the ante bellum years, although it was subject to fluctuations of as much as $10 a ton.[33] This is not surprising, considering on the one hand the lack of interest shown in the West toward the newer means of producing pig iron and, on the other hand, the expanding geographical basis of the industry which obviated a raw material limitation to an increase in production. In the East, pig iron prices showed a slight decline due to the switch to anthracite as a blast-furnace fuel in that area. However, the prices for each separate grade of iron (they were then graded according to fuel used) did not decline, and the price differential between these two grades was not severe: about $2 a ton on a price of about $30.[34]

The price of bar iron acted in a different manner. In the eighteenth century, the price of bar iron was three or more times the price of pig.[35] But at least partly due to the introduction of puddling and rolling, the price of bar iron began to decline before that of pig iron. By the 1840's when puddled bar appeared regularly in market quotations, its price was only two-thirds that of hammered bar. In addition, the prices of both hammered and puddled bar declined steadily in the half-century preceding the Civil War, from three times as expensive as pig iron or more, to twice as expensive or less. Hammered bar declined more than puddled bar as the quality of puddled

[32] The Warren and Pearson price index is meant. U.S. Bureau of the Census, *Historical Statistics of the United States, Colonial Times to 1957* (Washington, D. C., 1960), p. 115.

[33] Thomas Senior Berry, *Western Prices Before 1861* (Cambridge: Harvard University Press, 1943), p. 264; Louis C. Hunter, "A Study of the Iron Industry at Pittsburgh before 1860," unpublished doctoral dissertation, Harvard University, 1928, pp. 77, 392.

[34] Appendix C, Table C.15.

[35] A. C. Bining, *The Pennsylvania Iron Manufacture in the Eighteenth Century* (Harrisburg: Pennsylvania Historical Commission, 1938), pp. 177-178; Charles S. Boyer, *Early Forges and Furnaces in New Jersey* (Philadelphia: University of Pennsylvania Press, 1931), pp. 10-11.

bar increased, forcing hammered bar to match its price. In 1856 the price of the large items made of wrought iron — rails, bars, rods — was just about double that of foundry pig iron; the price of more specialized and quantitatively less important products such as boiler and sheet iron and hammered shapes had remained closer to four times that of pig iron.[36] Price statistics are available for a selected list of products from the 1850's on. They show the price of best refined rolled iron bars to have been almost three times that of pig iron, and the price of iron rails to have been 2.3 times that of pig iron in 1850. In 1870, these ratios had fallen to 2.4 and 2.2, respectively.[37] The price of wrought-iron articles thus continued to fall relative to that of pig iron after 1850, although the rate of descent may have declined.

At the beginning of our period, bar iron was a semiproduct that needed to be worked into some more specific shape to be used, just as pig iron for castings needed to be cast into shape. A comparison of prices between the two is justifiable on these grounds. By the middle of the nineteenth century, however, a substantial proportion of wrought iron was being sold in finished shapes, such as rails, that did not need further treatment to be used. The price of wrought iron in these forms should properly be compared with that of castings, not that of pig iron. This has not been done as there are no adequate statistics on the price of castings. What data there are, however, indicate that prices for castings were quite inflexible in the ante bellum years, which means that they did not have a trend relative to the also trendless pig iron price.[38] This is supported by the absence of any information suggesting changes in foundry practice that would lead to an alteration in the ratio of the prices of pig iron and iron castings. The data presented above on the prices of wrought and cast iron as they have been recorded by contemporary observers may thus be used to help discriminate between changes in supply and demand curves for wrought iron and iron castings.

[36] Berry, pp. 259-264; J. P. Lesley, *The Iron Manufacturer's Guide to the Furnaces, Forges and Rolling Mills of the United States* (New York, 1859), pp. 763-764.

[37] Appendix C, Table C.15.

[38] Berry, p. 276. To the extent that the degree of fabrication of wrought-iron products increased over this period, the price decline for wrought-iron is underestimated.

The other information necessary to discriminate between these two types of changes is contained in the following propositions.[39] The supply curves of both wrought and cast iron were quite elastic, if a few years — say two — are allowed for new firms to enter the industry. The industry was composed of a great number of small firms, each exploiting the raw materials locally available with the aid of skills that were widely known. As it was quite easy for iron production of both types to be started or discontinued and the unit of production was small relative to the total volume of production, the industry approximated the conditions of constant costs and had very elastic supply curves. The demand for wrought and cast iron was undoubtedly more sensitive to prices. The level of prices certainly influenced the nature and extent in which these materials were used; their demand curves, in other words, were relatively inelastic. In precise terms, we assume that the elasticities of the supply and demand curves in the markets for cast and wrought iron were similar, and that the supply curves were infinitely price-elastic while the demand curves were much less so.

If the supply curves for the two materials were infinitely elastic, the change in their relative prices must have been the result of a shift in the relationship of the supply curves alone. In other words, the fall in the price of wrought iron relative to that of cast iron indicates that the supply curve for wrought iron was falling faster than the supply curve for cast iron. And if the change in relative prices was due to the movements of the supply curves alone, the change in the relative quantities consumed must have been due to the movements of the demand curves alone. The rise in the quantity of castings consumed relative to the quantity of wrought iron consumed between 1830 and 1850, therefore, was due to a faster rise in the demand for iron castings than for wrought iron. The supply curve for wrought iron was falling faster than the supply curve for cast iron at the same time that the demand curve for cast iron was rising faster than the demand curve for wrought iron.

This opposition of forces was corrected after 1850, and the quantity of cast iron consumed no longer continued to rise faster

[39] This way of resolving the identification problem is of course only one way among many. It seems the most appropriate given the conditions in the iron industry at the time and the nature of the extant data, but other procedures would also resolve the ambiguities in the price and quantity data — possibly with other conclusions.

than the volume of wrought iron. The supply curve of wrought iron continued to fall faster than the supply curve for castings, as is shown by the movements of relative prices, but the demand curve for castings was no longer rising faster than the demand curve for wrought iron. In this fashion the trend away from the use of iron in the form of castings that was continued and extended after the Civil War was started.[40]

The reasoning used above applies *a fortiori* to the shifts in the composition of production that were noted above, the only difference being that the curves now refer to the supply and demand of goods produced in the United States rather than goods consumed in the United States. (In a competitive market, the prices are the same.) The movement of production toward castings in the years before 1850 was stronger than the movement in consumption because the rise in demand for wrought-iron products was more easily satisfied by imports than the rise in demand for castings, despite a fall in the domestic supply curve of wrought iron that resulted from the introduction of puddling and rolling.

This brief era when cast iron was the leading component of the iron industry shows the effects of lags in the economic system. For it was the product of a temporary opposition in the direction of movement of the supply and demand for iron products, or, to state it more clearly, of a temporary contradiction between the increasing ease of making wrought iron and the increasing uses for cast iron. Speedily corrected, this era may appear as an aberration in the course of development. To correct that impression, it may be noted that an analogous rise in the use of castings occurred in England, although it was somewhat earlier and slower. It began with the introduction of pig iron made with coke, cheaper than the older types of iron, but superior for the foundry and inferior for making bar iron. The early phase of iron machinery and artifacts was built upon the use of cast iron, a phase which did not end until the operations of rolling mills had been substantially improved

[40] The complete separation of the effects of shifting supply and demand curves used in the above argument depends on the infinite elasticity of the supply curves for these products. To the extent that the elasticities were less than infinite, the conclusions about the movements of the supply curves must be qualified. The conclusions about the movement of the demand curves before 1850, however, can be derived from the assumption of similar elasticities in the two markets alone.

and events such as the introduction of railroads had changed the pattern of demand.[41]

Somewhat similar conditions produced the American age of cast iron, although this age was of shorter duration than the comparable one in England. The American economy was changing rapidly, and the new uses for iron followed quickly upon each other. By the time the use of cast iron had become widespread, in the 1840's, new uses of iron and new ways of treating wrought iron had been introduced. As they gained in importance, they combined with the increasing relative ease of making wrought iron to shift consumption away from iron castings.

The following chapter details the movements in the demand for iron as it swung toward the increasing use of castings before 1850 and away from the use of that form afterward. Chapter 3 explores the reasons for the increasing relative ease of producing wrought iron, which followed from the slow rate of technological change in the production of pig iron.

[41] Ashton, pp. 38-40, 140-141; Gloag and Bridgwater, pp. 44-49, 53-54.

2

The Pattern of Demand

Before 1850

The previous chapter has shown that the changing composition of iron consumption in the years preceding 1850 was primarily a result of changes in the pattern of demand. In 1830, about four-fifths of the iron consumed in final form (measured by weight) was wrought iron. This proportion fell to about three-fifths in 1850, and shortly thereafter it began to rise. The composition of production followed the path of consumption in exaggerated form, due to the nature of imports, and the importance of the changing pattern of demand was magnified. The problems of production will be considered in detail in succeeding chapters; this chapter is concerned with the pattern of demand.

In order to make clear just what demand is meant, it is necessary to specify who is being discussed. This history is concerned with the makers of crude iron, rather than the makers of goods in final demand. Therefore, although the problems of iron founders are considered, they appear as generators of demand for pig iron to use in castings, rather than as creators of the supply of castings. In the fabrication of wrought-iron products there was an additional step, and the firms that refined pig iron into wrought iron were still the makers of intermediate goods. They are consequently considered as suppliers in our scheme, even though certain of their heavy, standardized products underwent no further processing after they left their mills. The sources of demand for wrought iron were the people who used iron products as intermediate goods, whether or not this implied further processing of the iron. (A railroad, for instance,

35

did not process the rails it used, but a blacksmith had to shape the iron bars he bought for his specific uses.)

The colonial iron industry satisfied a number of demands, none of which dominated the industry. The diversity of these demands can be seen in the list of iron articles produced in the colonial era given by one recent author:

> nails, spikes, brads, tacks, and wire; shovels, spades, hammers, and hoes; horseshoes, bits, and stirrups; edged tools (axes, chisels, handsaws, knives, scythes, mill saws); anvils, anchors, and cannon; stoves, skillets, kettles, and other hollow ware; and the iron parts of mill machinery, engines, firearms, wagons, harrows, and plows.[1]

These products may be grouped into five categories for the purposes of the following discussion. We are attempting to explain the shifts in demand in the pre-Civil War era between wrought and cast iron, and the categories have been selected for the purpose of furthering this particular inquiry. Unfortunately, the extant quantitative data do not allow us to assess the importance of these groups with any accuracy.

Household goods comprise the first category. These items — stoves, hollow ware, and miscellaneous implements of similar character — were made primarily of cast iron; stoves formed one of the identifiable large demands for this metal. Household goods accounted for most of the castings made in the American colonies, but rapid growth in the demand for castings did not come from this sector. Without being overly precise, household products were probably produced in some rough correspondence to the growth of population and are, therefore, not of large moment in explaining the growth of per-capita production.

The tools of agriculture can be separated into a category of their own, as can the iron components of machinery. These two classes used both cast and wrought iron, and it is not possible to tell precisely how much of each. In agriculture before the general introduction of the cast-iron plow, the main iron implements were edge tools and those hand tools that had been constructed or reinforced with iron by a local blacksmith or forge. Machinery for nonagricultural uses was similar, with the addition of cast-iron anvils and firearms of varied sizes.

---

[1] Curtis P. Nettels, *The Emergence of a National Economy* (New York: Holt, Rinehart and Winston, 1962), p. 270.

Cannon at this time were made of cast iron, but their quantitative importance is not clear.

Iron products used in construction form a fourth category. In the colonial period and the early years of the new republic, nails were the main use of iron in structures. These nails were wrought or slit nails, square in shape and made out of the first rolled iron plate. In the colonial period almost the only rolled iron products were the products of rolling and slitting mills, which produced either nail plate or strips of this plate for the fabrication of nails. Iron used for transportation equipment, the last category, was a mixture of cast and wrought iron. The main products at this time were fittings for horses, horse-drawn wagons, and ships, different parts of which were made of different materials.

The products in the list quoted above can be fitted into one or another of these categories. To which one should we look for the expansion of the demand for cast iron at the expense of wrought before 1850? Where should we expect to see the effect of the cheapened supply of wrought iron first manifest itself in an interruption of the movement toward castings? The shift in the demand curves toward cast iron was a general, if mild, phenomenon; more or less ambiguous evidence on it can be found in all categories save the first, which started out being mostly castings. But when we look for the opposing trend after 1850, the last two categories, and particularly transportation equipment, become more important than the other sectors.

In addition to the changing product structure to be discussed shortly, two general influences contributing to the demand for cast iron may be noticed. They are the beginning use of mineral fuel in the blast furnace, and advances in foundry technique that enabled the foundries to supply a wide range of products.

In Britain, the primary stimulus for the use of castings came from the introduction of coke as a blast-furnace fuel. The pig iron produced was a fine foundry metal, due to its increased fluidity, but an inferior forge material.[2] In the United States, coke was not of great importance in this period, and anthracite was adopted only in one part of the country. Nevertheless, anthracite pig iron shared the characteristics of coke pig iron, albeit in lesser degree, and was an influence in the same di-

[2] See Chapter 1.

rection. American ironmasters used American anthracite pig for castings, and commented on its difficulties in other uses.[3]

In addition to the change in the quality of iron used, the foundry of the 1830's and 1840's was better able to supply a variety of needs than the forges. In the foundry, the metal was shaped by pouring it into a mold made of dampened sand; in the forge, it was shaped by altering its position under successive blows of a trip hammer. The foundry obviously offered greater possibilities for the fabrication of intricately shaped pieces, and it was also able to make larger and stronger pieces than was the forge. In addition, the foundries separated from the blast furnaces at this time and associated themselves more closely with their consumers. By so doing, they rendered themselves more responsive to the demand exercised by their customers and more flexible with regard to their location. The most common type of business for a foundry to be associated with was an engine works; one-fourth of the thirty-eight foundries in Pittsburgh in 1854 made steam engines exclusively.[4]

This last development was aided by the use of mineral fuel, for the use of a separate foundry depended on the ability of the founder to remelt the pig iron in an economical way. The alternative to casting directly from the blast furnace benefited from the introduction of the cupola furnace, essentially a smaller blast furnace, for remelting the iron. The use of the cupola became widespread in America when anthracite was introduced, making anthracite a progenitor of iron castings in two ways. It was adopted in England after 1795, and improved by the American innovation of the drop-bottom about 1850.[5]

Turning to specific products, we can ascertain the size of the first category from the Census. The stove industry split off from other types of foundries and became a separate industry during

[3] Trenton Iron Company, *Documents Relating to the Trenton Iron Company* (New York, 1854), p. 4; George H. Thurston, *Pittsburgh as It Is* (Pittsburgh, 1857), p. 103. Thurston also notices that the limited supplies of coke iron are excellent for castings.

[4] R. N. Grosse, "Determinants of the Size of Iron and Steel Firms in the United States, 1820-1880," unpublished doctoral dissertation, Harvard University, 1948, pp. 33-35; Victor S. Clark, *History of Manufactures in the United States* (3 vols.; New York: McGraw-Hill, 1929), Vol. I, pp. 502-503; *ARJ*, 27 (1854), 521. The full titles of frequently cited periodicals can be found in the Bibliography.

[5] Henry Jeffers Noble, *History of the Cast Iron Pressure Pipe Industry in the United States of America* (New York: Newcomen Society, 1940), p. 31; Clark, Vol. I, p. 416; John Gloag and Derek Bridgwater, *A History of Cast Iron in Architecture* (London: Allen and Unwin, 1948), pp. 48, 63.

this period, and it is reported separately from other castings in the Census of 1860. It had a product valued at $11 million out of a total value of $36 million for all castings.[6] In other words, about 30 per cent of all castings were produced by establishments whose primary product was stoves, although an admixture of other household goods was undoubtedly included in this proportion.

The second category, agricultural implements, underwent a significant change in the years under discussion. The two dominant innovations were the cast-iron plow and the horse-drawn reaper, although many other kinds of farm machinery were introduced in this period. The first cast-iron plow to win people away from wooden plows was patented by Jethro Wood in 1814 and 1819. This plow was distinguished by having a mold board of cast iron with a replaceable edge. Casting a plow in several pieces to facilitate cheap replacement of worn-out or broken parts was the crucial factor in winning acceptance for the cast-iron plow, and by 1830 it had come into general use in New England and the Middle Atlantic states; it never made serious inroads into the West due to its inability to cut the tough prairie soils.[7]

Further improvements in the plow followed, leading to the establishment of many plow-making firms and to the later introduction of the cast-steel plow. Progress was duplicated in other areas of agricultural activity, and the process of substituting iron implements for wooden, fast machines for slow, horse power for human, went forward. Perhaps the outstanding example, certainly the most well-known, was the introduction of the reaper by McCormick and Hussey. The date of invention is usually given as 1834, but there were very few reapers in use by 1845. McCormick moved to Chicago shortly thereafter and the production of reapers rose; most of the wheat grown in

[6] U.S. Census, Eighth, III, *Manufactures of the United States in 1860* (Washington, 1865), p. clxxxvii. The Census also notes that the value of castings other than stoves did not rise between 1850 and 1860 (p. clxxxvi). In light of the static price level, this implies that the quantity did not rise, and supports the evidence of stagnation given in Chapter 1.

[7] Paul W. Gates, *The Farmer's Age: Agriculture, 1815-1860* (New York: Holt, Rinehart and Winston, 1960), pp. 279-281; Leo Rogin, *The Introduction of Farm Machinery in its Relation to the Productivity of Labor in the Agriculture of the United States During the Nineteenth Century*, University of California Publications in Economics, Vol. 9 (Berkeley: University of California Press, 1931), pp. 21-31.

Ohio and on the prairies was cut by machine at the end of the ante bellum period. Conditions during the Civil War gave tremendous impetus to the trend to mechanization, and "more harvesting machines were produced in the few years of the Civil War than had been turned out during the entire period which elapsed from the time Hussey sold his first machine in 1833 to the outbreak of the struggle." The total number of reaping and mowing machines produced and in use by 1864 was on the order of 250,000.[8]

What was the significance of these and the other similar developments in farm machinery for the production of iron? They increased the use of cast iron, obviously in the case of the plow, but also in the many cast parts of the more complex machines. The larger machines, however, also required large quantities of wrought iron, and it is not known how much these developments altered the balance between cast and wrought iron. The number of cast-iron plows in use by the end of the Civil War is not known, but the small number of reapers that had been produced indicates that the amount of iron used in their production was probably small relative to the amount of iron produced. The combined influence of other agricultural machinery on the production of iron was also not large, as most of the value of these machines was in the work done on the iron rather than in the iron itself.

The field of nonagricultural machinery was dominated by one type of machine in the years preceding the Civil War: the steam engine. Table 2.1 indicates the share of total machinery reported by different classes in the 1860 Census.

Steam engines accounted for most of the machines built in this period, although a value calculation probably overstates the proportion of iron used by steam engines, since the value added to each ton of iron used was probably higher for steam engines than for other machines. But the proportion of value embodied in steam engines was so high as to make any necessary modification to the figures of secondary importance.

The steam engine needed iron for its existence and used cast iron for many of its parts. As it replaced water as the prime mover in many branches of industry, it increased the consumption of cast iron. A Congressional report in 1838 estimated that there were 3,010 steam engines in the United States, of which they

[8] Rogin, pp. 72-93 (quote, p. 93).

TABLE 2.1
DISTRIBUTION OF THE VALUE
OF MACHINERY PRODUCED, 1860
(percentages)

Steam engines	90.0
Cotton and woolen	9.4
Woodworking	0.3
Turbine water wheel	0.2
Other	0.2

Source: Eighth Census, p. 738. This sector represents less than 3 per cent of the value of manufacturing production reported by the Census.

had actually observed 2,653. Of these engines, 1,860 were stationary engines and therefore, except for a relatively small number on the Pennsylvania Portage Railroad and similar projects, in industrial establishments. Of the remainder, 350 were in locomotives and 800 in steamboats. Pennsylvania and New York had the largest number of steamboats, and Pennsylvania had the largest number of locomotives, showing that powered movement was still primarily a possession of the East. Pennsylvania also led in the number of stationary engines, followed by Louisiana, Massachusetts, Virginia, and New York. Industry had yet to move away from the entrepots of water transport, where the rivers came to the sea.[9] The effects of all this, however, on the production of crude iron were again only partial. Most of the value of steam engines, as of horse-drawn agricultural machinery, was the result of the labor necessary for their construction, not of the raw materials used.

For the remainder of machinery, in which textile machinery dominated as much as steam engines in the larger group, the trend toward the use of cast iron was also strong. The main impetus to the use of iron in machinery was given by the new steam engines, but the signs of the switch were evident elsewhere. Clark summarizes some of these developments:

Following the substitution of iron for wood in steam-engines came its employment in other large machinery. Though as late as 1840 the framing of American textile machinery was frequently of wood,

[9] U.S., Congress, House, Executive Document 21, 25th Cong., 3rd Sess., "Report on the Steam Engines in the United States" (1838); Clark, Vol. I, pp. 505-510.

the substitution of cast iron for this material dated from the introduction of power-looms. In 1820 Worcester and Fall River founders were making castings for the latter, which sometimes were designated as iron-side looms. Two years later a Worcester maker began to cast cylinders for cards, making them in four parallel pieces. With the introduction of turbines came a demand for cast iron pipes, shafts, and casings. It is said that improvements in casting were "contemporaneous with and made possible this new design of waterwheel." For many years castings were used for machine parts for which forgings later were employed.[10]

The use of iron for machines formerly made of wood, reached as far as the machinery for making the iron itself; in the 1830's cast-iron cylinders began to replace the wooden tubs that had previously pumped air into the blast furnace. And in the areas of woodworking and miscellaneous machinery, where American technology was probably as advanced vis-a-vis Britain as in any other sector, the trend toward iron was also apparent, although British observers who visited America noticed that this process was not as far advanced here as there.[11]

Construction and transportation, two sectors that were closely related, shared in the general shift toward cast iron and supplied the primary impulses for the reversal of the trend in demand. It was in these sectors that the introduction of new products, such as structural iron, steamboats, and railroads, altered the face of our country and the type of iron demanded. The products involved were heavy and standardized, and they were produced in bulk. And while the quantity of the castings made for them is unknown and therefore cannot be accurately compared with other demands for cast iron, the amount of wrought iron going to the major use in this category, rails, is known with some accuracy, and its effect in turning the composition of iron production back toward wrought iron is easily seen.

Within construction three products are worthy of notice: nails, pipes, and structural iron. Nails were made, throughout the life of the iron nail, of wrought iron. They were very scarce in colonial days, houses often being burnt down when abandoned to recover the nails. But in the early decades of the nineteenth century, nail machines were introduced that speeded their

---

[10] Clark, Vol. I, p. 416.

[11] B. F. French, *Rise and Progress of the Iron Trade of the United States from 1621 to 1857* (New York, 1858), pp. 55-56; D. L. Burn, "The Genesis of American Engineering Competition, 1850-1870," *Economic History*, A Supplement to the *Economic Journal*, 2 (1931), 292-311.

manufacture and lowered their price. Machines had been patented for cutting the nail plate as well as for making nails out of the resulting strips. They were widely adopted despite the difficulty of keeping them in repair.[12]

Cast-iron pipe had its main use in municipal water and gas systems, the former being the more important. The original water systems introduced into a few Eastern cities in the opening years of the nineteenth century used wooden pipes made of bored-out logs. Philadelphia was the first city to start substituting iron for wood, in 1820; its first pipes were obtained from England. New York, Baltimore, and Boston followed in the following two decades, using both British and American pipe. Pipes, like other cast iron articles, could be cast directly from the blast furnace or made from remelted iron. It is not known which method was the one more often used, although the first cast-iron pipes made in this country (about 1830, for the Philadelphia Water Works) were cast directly from the blast furnace.[13]

Cast iron had begun to be used in bridges in England by the end of the eighteenth century, one of the earliest designs being supplied by an American. The design was created by Thomas Paine for the purpose of spanning the Schuykill River, but he went to England, and the bridge was placed on a bowling green at Paddington to be exhibited to the public for a shilling a head. The outbreak of the French Revolution diverted Paine's attention from this structure, but in a modified form it was erected over the River Wear in 1796.[14]

The use of cast iron for bridges, and later for buildings, spread rapidly in England and somewhat less rapidly in America. Wrought-iron bolts were used as tension members in wooden bridges; cast iron replaced masonry arches in larger structures. The form of the older stone structures was adopted in lighter form when cast iron was used, the similar strengths of the two materials in resisting compression producing such a development. Around 1840 in America the two types of iron were used together as bridges made entirely of iron replaced the bridges

[12] James M. Swank, *History of the Manufacture of Iron in All Ages* (second edition; Philadelphia, 1892), pp. 448-450; Clark, Vol. I, pp. 222, 516.

[13] Nelson Manfred Blake, *Water for the Cities* (Syracuse, New York: Syracuse University Press, 1956), pp. 83, 122, 162-163, 190, 220; *Bulletin, 31* (1897), 50.

[14] Thomas S. Ashton, *Iron and Steel in the Industrial Revolution* (Manchester: At the University Press, 1924), p. 141; *Hunt's, 43* (1860), 498-499; Gloag and Bridgwater, p. 86.

made of iron and wood. In the construction of buildings, however, cast iron only was used, forming rafters, pillars, and store fronts, and creating the "fireproof" building. Cast iron could dominate building construction more easily than bridge construction due to the more static nature of the stresses to be undergone, and in this area of construction technology, America probably led England a little.[15]

The use of cast iron in building construction came into its own about the middle of the nineteenth century, and some of the best examples of this type of construction were built in St. Louis. In this city an entire quarter was constructed in the course of the 1850's using cast-iron and glass construction. The buildings have a very "modern" look about them, and they illustrate how sophisticated the users of cast iron had become by 1850.[16]

After 1850

The shift toward cast iron in the years before 1850 may be seen as the result of general forces acting on many products. These influences were the ease with which cast iron could be shaped, its greater rigidity in small machine parts, its greater strength in larger structural items, and, of course, the nature of the products in demand.

But cast iron was not the universal material, and the demands of the economy changed. In addition, the ability of the British iron industry, and later the American iron industry also, to supply cheaply the desired articles of wrought iron altered the pattern of consumption. The trend toward cast iron was halted before it had progressed very far, and an opposing trend set in. The products which showed the progress of wrought iron earliest were rails and structural iron. Although rails led these two, the discussion of iron for construction may profitably be finished before examining rails.

As the ability to make large, reliable wrought-iron girders and pieces increased, and as wrought iron became cheaper relative to cast iron, people switched to this type of iron for a con-

[15] Theodore Cooper, "The Use of Steel for Bridges," *Transactions of the American Society of Civil Engineers*, 8 (1879), 263-277; ARJ, 27 (1854), 747, 779; Clark, Vol. I, p. 504; Gloag and Bridgwater, p. 198.

[16] Sigfried Giedion, *Space, Time and Architecture* (fourth edition; Cambridge: Harvard University Press, 1962), pp. 165-206.

struction material. The bridges built during and near the time of the Civil War appear to have been a combination of cast and wrought iron, mostly the former, but with the proportion of the latter rising; buildings of this time were still based upon cast iron, but the shape of things to come was foreshadowed by the construction of Cooper Union in New York with wrought-iron beams in the 1850's.[17]

Although cast iron was not to be deprived of a major role in the supporting parts of construction until the advent of cheap steel, the considerations that favored the use of wrought iron before the age of steel had a wide influence. One view of the relative merits of the two types of iron can be seen in a comment by an English observer in 1877:

> In the modern construction of girders and bridges, cast iron has, to a great extent, been superseded by wrought iron, because of the superiority which the latter possesses over the former against every kind of strain (tensile, transverse, shearing, vibrating strains, and those caused by the effects of temperature), with the exception of the quiescent compressive strain. Cast iron is much stronger than wrought iron in resistance to compression; and, owing to the existence of a natural crust or skin on its surface, it is not so liable as wrought iron to deterioration by rust, and, if kept painted, with ordinary care will prove the more enduring of the two. In the facility which it possesses of taking all possible shapes—whether required by a minute regard to the forces in a structure or for purposes of ornament—cast iron is also superior to wrought iron; while, weight for weight, it is much cheaper . . . [But] it cannot be wondered at that failures arising from the improper application of cast iron, or the belief that all cast iron is of uniform quality, should have deterred many people from using it at all in bridge and girder construction.[18]

Within the transportation sector, major changes were in progress. According to Hunter, "What the railroad was to the iron industry after 1850, the steamboat had been in large measure during the preceding generation."[19] The steamboat had a wooden hull, but used an iron power plant and accessories. In addi-

---

[17] Andrew Carnegie, *Autobiography* (Boston: Houghton Mifflin, 1920), pp. 111, 125; *Hunt's*, 45 (1861), 291; Allan Nevins, *Abram S. Hewitt* (New York: Harper and Brothers, 1935), pp. 114-115.

[18] Ewing Matheson, *Works in Iron* (London, 1877), p. 52, quoted in Gloag and Bridgwater, pp. 247-248.

[19] Louis C. Hunter, "The Heavy Industries Before 1860," in Harold F. Williamson (editor), *The Growth of the American Economy* (New York: Prentice-Hall, 1944), p. 181.

tion to the power plant, the railroad used metal rails, and when it was decided that they would be made of wrought iron, the balance was tipped toward this material in the composition of consumption. The railroads represented a marginal shift; neither the steamboat nor the railroad accounted for most of the iron produced before the end of the Civil War. Nevertheless, it was here that the consumption of wrought iron began to grow faster than the consumption of cast iron, and it was here that the change was first visible.

By 1856, at the end of the railway boom, the products of wrought iron could be represented by Table 2.2 showing the distribution of consumption, production, and imports. It should be remembered that the production of wrought iron in the United States supplied less than two-thirds of the consumption.

TABLE 2.2
COMPOSITION OF WROUGHT-IRON CONSUMPTION,
PRODUCTION AND IMPORTS, 1856
(percentages)

	Consumption	Production	Imports
Rails	38	27	56
Bars and rods	40	45	35
Nails	10	16	0
Boiler and sheet iron	7	7	5
Hammered shapes	4	4	4

Source: J. P. Lesley, *The Iron Manufacturer's Guide to the Furnaces, Forges and Rolling Mills of the United States* (New York, 1859), pp. 763-764. The imports of bars and hammered shapes were not separated by Lesley: I took the liberty of separating them according to the composition of domestic production. This cannot be too far wrong, as Lesley's combined imports are less than 20 per cent higher than the imports of bars alone. (Appendix C, Table C.13.) Totals of columns may not add to 100 due to rounding.

The dominant new product of the age was of course rails, which had risen from being inconsequential in 1830 to over one-third of the total consumption in 1856. The proportion of production was closer to one-fourth, as over half the rails used in this country were supplied from England. That this product was not more important for the American iron industry cannot, in the light of these data, be laid at the door of demand. The major part, possibly the overwhelming part, of the increase in the consumption of wrought iron relative to cast iron came from rails. Bars, nails, and boiler iron had been consumed before 1850, and there is no evidence that their quantity rose sharply. It was the cheapened supply of wrought iron, coupled with a rising demand for rails, that turned the tide.

The last, small category is worthy of more attention than its size might imply. In 1830, almost all the wrought iron made was hammered, and it was hammered shapes that offered competition to castings. But the production of wrought iron lagged until rolling was adopted. By 1856 the proportion of wrought iron that was shaped by hammers had fallen to an almost insignificant extent. This may be said despite the knowledge that bars could be made by either hammering or rolling, for the advent of grooved rails forced hammers from this field by the cheapness with which they enabled bars to be produced.

There were many questions that had to be settled in the course of the introduction of rails. It was quickly found that a material stronger than wood was needed, and that this material was likely to be iron. Then came the decision that is the theme of this chapter: wrought or cast iron. In addition, the weight and best shape for rails and the best foundation for the rails had to be found. These problems were vital issues in the 1840's, but by the 1850's their solutions had assumed their final form. The product to be used was decided early, for the qualities that rails needed were resilience and ability to absorb shock. In addition, when it was decided not to support the rails at every point, tensile strength was needed. The decision therefore went for wrought iron, although experiments involving the production and use of cast-iron rails continued into the 1840's.[20]

The shape of the rail proved problematical. At first all rails were bars laid on wooden sleepers; according to one calculation, 4,000 miles of railway were laid with such rails by 1845.[21] But these rails came loose, their ends curling to form "snakes' heads" that often pierced the floors of railroad cars. To avoid this difficulty, the iron rail was made heavier, and it assumed the function of holding the weight of the train in addition to providing a hard surface. In 1845, when the above calculation was made, the *American Railroad Journal* complained: "The American iron-masters appear to consider railroad iron as unworthy of their notice. We have understood from pretty good

[20] Swank, pp. 431-433.
[21] *Hunt's, 12* (1845), 67.

authority that not a bar of T rail has yet been rolled in the three great anthracite and iron districts of Pennsylvania!" By 1852 the *Journal* could note that "edge rails" were in common use, both the T or H rail and the inverted U or bridge rail. Each of these rails had a broad base supporting itself; the English practice of supporting the rail at each bearing by a "chair" had been abandoned, and the English I rail that was, in theory, reversible had never been used. The former was found to be more expensive than broadening the base of the rail sufficiently to enable it to support itself, and the latter had the disadvantage of providing a rail that when reversed was nicked at each point at which it had been supported, often deeply enough to be unworkable.[22]

A long debate ensued on the transition to edge rails, the question being the location of production of the first T rail to be made in the United States. (The bridge rail was abandoned after a short delay, and its origins were never hotly contested.) The honors appear to belong to the Mount Savage Rolling Mill, of Allegheny County, Maryland. In 1844, they rolled T rails for the railroad from Fall River to Boston. (They were also, it is interesting to note, one of the original users of coke in a blast furnace in this country.) But the significance of the location of the first T rail production should not be exaggerated; within two years, there were at least six other establishments rolling T rails, and more followed.[23]

The weight of rails did not follow immediately from the decision to use shaped rather than flat rails, or even T rails rather than others. T rails were made in weights ranging from 25 to 68 pounds to the yard by one firm alone, and in higher weights by others. The optimum weight of this period was seen to be 60 to 65 pounds to the yard by contemporary experts.[24]

The question of support for rails remained open to speculation for many years. The possibilities were cast iron, stone, and wood. The last two were the more important, and the last was the final solution due to its ability to absorb part of the shock from the impact of the locomotive wheel against the rail. The greater lasting ability of rails placed on wooden ties was a

[22] Swank, p. 429; *ARJ, 18* (1845), 265; *25* (1852), 130; Zerah Colburn and Alexander L. Holley, *The Permanent Way and Coal-Burning Locomotive Boilers of European Railways* (New York, 1858), p. 78.

[23] Swank, pp. 433-435; Chapter 3.

[24] *Bulletin, 4* (1869), 45; Colburn and Holley, p. 80.

decisive argument in their favor, even though some iron ties were still in use as late as 1877.[25]

This, of course, was the big difficulty with iron rails – they wore out. Or, more precisely, "Rails seldom *wear* out – they laminate or crush in the majority of cases."[26] The latter difficulty was the result of sudden shocks which could be only partially eliminated by the use of elastic ties. The former problem derived from the method of manufacturing iron rails. They were not rolled from one large bar which had been converted to wrought iron as a unit, because the puddling process did not allow large units to be handled at a time. They were built up from a number of smaller bars placed together to form a "pile" which was heated to welding heat and rolled into the shape of a rail. The weld was never complete, and the strips that made up the rail would come apart after more or less use, that is, the rail would laminate. This difficulty was never satisfactorily solved with the wrought-iron rail, although American rail makers learned to control it sufficiently to compete with English manufacturers. The lack of durability of iron rails formed one of the chief stimuli to the development of cheap steel and will be explored more fully under that heading.

Here we may note that as rails wore out, they were available to the market as scrap iron. Because of this, the consumption of pig iron for use in rails grew slower than the production of rails, even though the production grew slower than the consumption of rails. This turn of events was the result of the discovery that old rails could be rerolled into new ones; they were heated and rolled into bars, then combined (usually with some new wrought iron) into the "pile" from which the new rail was made. By the time of the Civil War, about half of the rails produced were rerolled rails, mostly in the West.[27] This meant that the significance of rails for the producers of pig iron was less than its significance for the producers of wrought iron in terms of proportion of product.[28] In terms of opportunities lost in the 1850's, however, it would be harder to make the

[25] *ARJ, 18* (1845), 649-651; Carnegie, p. 88; *Bulletin, 11* (1877), 141.
[26] Colburn and Holley, p. 88.
[27] Samuel Harries Daddow and Benjamin Bannan, *Coal, Iron and Oil, or the Practical American Miner* (Pottsville, Pennsylvania, 1866), pp. 694-695; Chapter 5.
[28] See Robert W. Fogel, "Railroads and American Economic Growth: Essays in Econometric History," unpublished doctoral dissertation, Johns Hopkins, 1963, Chapter V.

equivalent statement, in view of the static nature of pig iron production in that decade.

This chapter might well close upon that note, but there is another demand which must be mentioned: the demand for steel. There were many iron products that have not been discussed here, such as wire and plates for ships and boilers, because they were quantitatively small in this period. The amount of steel consumed (measured by weight) was even smaller, but it is of interest in connection with the later development of the industry. Steel was an expensive, specialized product that could only be used in small quantities or where considerations of quality dominated considerations of cost. In addition, the manufacture of good steel was very difficult, and very little high-quality steel was made in the United States before the Civil War.[29] The uses of steel, therefore, were restricted to edge tools, cutlery, prairie plows, and other products where small amounts of a durable material made a large difference.

During the Civil War, American railroads began to experiment with imported steel rails, admitting the possibility that even a vastly more expensive rail might be preferred if it would wear better than an iron rail. But the contest was not joined, for new methods of making steel had been discovered and were put into use shortly after the war. That is the second half of this history; we have been concerned here with the demand for iron before 1865, and relative to the demand for iron, the demand for steel in this period need only be briefly noted.

[29] Swank, pp. 383-394.

3

A Question of Fuel

The Neglect of Coke

The discussion of supply and demand at the end of Chapter 1 uncovered a fall in the supply curve for wrought iron relative to the supply curve for pig iron in the years prior to 1865. The supply curves were changing under the influence of a new iron-making technology being imported from England, and the supply curve for wrought iron was falling faster than the supply curve for pig iron due to a greater eagerness on the part of wrought-iron producers to use the new technology. The great expansion of the iron industry's capacity that resulted from the boom of the 1840's offered a good opportunity for the new technology to be adopted; puddling and rolling were adopted throughout the industry, but the use of mineral fuel in the blast furnace was adopted in only one area. To understand the relative movements of the supply curves, it is necessary to discover why coke was ignored in the West as a blast-furnace fuel. We wish to know why Western pig iron producers ignored available innovations, and the question, as the title of this chapter indicates, is a question of fuel.

Although this problem has arisen in the context of the conflicting forces at work in the ante bellum iron industry, it has previously been of concern for other reasons. The slow spread of coke as a blast-furnace fuel from England, where it was adopted at the end of the eighteenth century, to other countries, who usually ignored it until the mid-nineteenth century, has raised questions about the nature of the diffusion of technology. From this point of view, it is of concern to find out what delayed the spread of the new technology and whether the obstacles

51

are the kind that are found elsewhere. From both points of view, it is of interest to ascertain whether the limits to the expansion of the new technology came from the supply side or, somewhat paradoxically, from the demand side.

Before the 1830's all the pig iron made in this country, with trivial exceptions, had been made with charcoal; in the 1830's, mineral coal began to be used for this purpose. There are two broad classes of mineral coal: bituminous coal, a soft coal from which the gas can be expelled (by coking), leaving a denser product, coke, more nearly approximating pure carbon; and anthracite, a hard coal containing very little gas in its natural state, and which has been repeatedly characterized as a "natural coke." Anthracite deposits in this country are highly concentrated in eastern Pennsylvania, and when mineral coal was adopted in the East, anthracite was used. Bituminous coal is found in many localities, but for our purposes most importantly in a region stretching across western Pennsylvania and Maryland, west of the Allegheny Mountains. When mineral fuel was employed in the West, bituminous coal was used, either in its raw form or as coke, the latter being the more suitable and the more important of the two forms.

Starting shortly before 1840, the production of pig iron with anthracite rose rapidly. By 1854, the year for which the American Iron Association first collected statistics, the pig iron made with anthracite amounted to over 45 per cent of the total pig iron produced in the United States. In this same year, pig iron made with bituminous coal in whatever form was only about 7.5 per cent of the total, the balance of the production being accounted for by pig iron still made with charcoal.[1] The new technology of the era, if we take that to mean the use of mineral fuel in the blast furnace, was then not ignored. It was only ignored in one variant and, what is the same thing, in one region. Interest in this problem has been generated from the fact that this ignored variant was the original English practice, while the variant used in America was only discovered close to 1840, and from the independent fact that the original English variant was the one in universal use by the end of the nineteenth century.[2]

[1] Appendix C, Table C.3.
[2] A variant here refers to the use of a different variety of fuel, and the differences in usage that followed from the differences in the coals.

The question, if posed in terms of the diffusion of technology, is why anthracite was first used in America when coke was the first to be used in England and was destined to become the one used in both countries after some delay. The same question, in terms of the lags discussed in previous chapters, is why the Western sector of the pig iron industry ignored the new technology, giving a slower rate of technological change to the smelting branch of the iron industry than to the refining branch which worked under no such difficulties. The identity of these questions is apparent from the geographical distribution of the industry and of coal deposits. As transportation was not well enough developed to allow iron or coal to be moved across the Allegheny Mountains in large quantities before the middle of the 1850's, the Western iron industry could use only bituminous coal or coke if it wanted a mineral fuel, and coke would be used only by Western ironmasters. The same is obviously true of Eastern ironmasters and anthracite.

The dual nature of the question emphasizes two possibilities for its answer. Demand may have been of the same type on the two sides of the mountains, only smaller in the West, while the supply curves were of different kinds. The difference in reactions would then have been due to the difference in supply and would have been corrected when the supply curves changed. Or demand may have been qualitatively different in the two regions, while the supply curves were similar aside from a difference in scale. The discrepancy among the regions in this case would have been eliminated as the result of changes in the nature of demand in one region. This is the traditional partitioning of a problem into considerations of changes in supply and of demand. It does not imply that either of them is autonomous, and the conclusion that, say, the supply curves were of different natures in the two regions would not mean that changes in the supply curves occurred autonomously. Any solution to this problem must link the changing technology of the West with the westward movement of the American economy; the question being asked here is the nature of the connection.

The existing literature does not indicate an answer to the question just posed. Most writers have considered the problem of why coke was not adopted without reference to what was happening in the East. Nineteenth-century writers attributed the continuing use of charcoal in the West to the ignorance or stupidity of ironmasters, i.e., to the nature of the supply curve of coke pig iron. This has been challenged in our century, but the issue has not been finally resolved. One of the earliest attempts by a historian of the industry to explain the delay in the adoption of coke appears in James M. Swank's book, *History of the Manufacture of Iron in All Ages*. He said the delay could be "variously explained," and did not attempt to choose among the many available possibilities, ranging from the plentiful supply of charcoal and the scarcity of suitable coal to the ignorance of ironmasters and the prejudice of iron buyers. These factors were taken over by Victor S. Clark in his *History of Manufactures in the United States*, who also declined to choose among these alternatives.[3]

F. W. Taussig, writing about the same time as Swank, gave his views on the subject in his *Tariff History of the United States:* "Charcoal iron for general use was a thing of the past [in the 1840's]; and the effect of the tariff of 1842 was to call into existence a number of furnaces which used antiquated methods, and before long must have been displaced in any event by anthracite furnaces."[4] This argument, stated in economic terms, says that the tariff of 1842 changed the price structure in such a way that the production of pig iron with charcoal was more profitable than it otherwise would have been. The imposition of the tariff raised the demand for domestically produced pig iron, and Taussig's argument asserts that at the higher price this produced in the market, the manufacture of charcoal pig iron was more profitable than at the lower price that would have ruled under free trade. If the prices of all types of pig iron rose uniformly, however, and if the manufacture of pig iron with coke was more profitable than the use of charcoal at a lower price, it was still more profitable at a higher one. The tariff might then *allow* ironmasters to use the older technology, but it would not *induce* them to. Only if the tariff altered the relative price of charcoal pig iron and coke pig iron (or of the inputs to pig iron manufacture) could the profitability of making pig iron with charcoal change from being less profitable than making pig iron with coke to being more profitable.

[3] James M. Swank, *History of the Manufacture of Iron in All Ages* (second edition; Philadelphia, 1892), p. 366; Victor S. Clark, *History of Manufactures in the United States* (3 vols.; New York: McGraw-Hill Book Company, Inc., 1929), Vol. I, p. 413.

[4] F. W. Taussig, *The Tariff History of the United States* (fourth edition; New York, 1898), p. 133.

A new argument was stated in 1929 by Louis C. Hunter in his classic article on the Pittsburgh iron industry.[5] He refuted the reasons given by Swank and used by Clark, and then went on to show why the lack of attention given to the use of coke in blast furnaces before 1860 was a reaction to existing market forces. He said that there was no impediment to the production of coke pig iron from the supply side, that coal, knowledge, and transportation were all available, but that the lack of demand for coke pig iron made its manufacture unprofitable. This, like Taussig's explanation, depends upon relative prices and the relative rates of profitability deriving from them. It differs from Taussig's explanation in attributing the price differential to the character of demand rather than to a (misguided) policy on international trade.

Hunter asserted that the failure of Western ironmasters to adopt coke as a blast-furnace fuel was a rational reaction to the market in existence at that time. The era could be characterized as an *agricultural* one in which iron was sold primarily as a semifinished product, bar iron, to be worked into final form by a local blacksmith or by the farmer himself. The ease of working such iron was more important than its cost, as it was easier to acquire the funds to pay for expensive iron than it was to acquire the skill to work with inferior iron. In addition, the iron had to be capable of maintaining high standards in a variety of uses, as it was not purchased for any specific one. The demand for iron, therefore, was much more sensitive to changes in quality than it was to changes in price.

About 1860 a change began to take place in the market, according to this argument, as it progressed toward one more typical of an *industrial* era. This newer market was typified by large-scale sales to purchasers buying for specific purposes. Each lot of iron had to satisfy only a few of the qualifications formerly imposed, and the sum of these individual demands produced a total demand that was less sensitive to differences in quality and more responsive to changes in price.

Charcoal pig iron could be transformed into bar iron that was easily worked and would give excellent service under a variety of conditions. Coke pig iron was substantially cheaper, but it possessed certain trace elements that gave it character-

[5] Louis C. Hunter, "The Influence of the Market upon Technique in the Iron Industry in Western Pennsylvania Up to 1860," *Journal of Economic and Business History*, 1 (February, 1929), 241-281.

istics harmful to this kind of general service. The most important of these trace elements was sulphur, which made the iron hot-short, that is, brittle when heated, and therefore difficult to work. As Overman, the author of a standard production manual of the time, put it: "We cannot pay too much attention to this subject [that is, sulphur], for upon it depends the success of the blast furnace operation. Sulphurous coal, by improper treatment, will produce sulphurous coke, and consequently sulphurous metal, which, in all subsequent manipulations, will be injurious, troublesome, and expensive."[6] Therefore, Hunter concluded, it was the influence of the market that first hindered and then encouraged the introduction of coke into the blast furnace.

Hunter was concerned only with the iron industry tributary to Pittsburgh in his investigations, and his explanation of technological change in the West was not formulated to account for the lag of the West behind the East. And since his argument talks in terms of both supply and demand, it may be extended in several ways. He maintained that coke pig iron was of a different quality from charcoal iron and only adopted when demand for this kind of iron rose around 1860. Anthracite could have been adopted earlier than coke because anthracite pig iron was more similar to charcoal iron than coke pig iron, or because the pattern of demand in the East was different than that in the West.[7] These correspond to the two possibilities listed above, and this brief survey of the literature has not answered the question at hand.[8]

Hunter's hypothesis, like Taussig's, depends on relative prices and on the assertion that it was not possible to make substantially greater profit in the production of pig iron with coke than

[6] Frederick Overman, *The Manufacture of Iron* (second edition; Philadelphia, 1854), p. 130.

[7] W. Paul Strassmann, *Risk and Technological Innovation* (Ithaca: Cornell University Press, 1959), pp. 24-25, discusses the introduction of both anthracite and coke, but he discusses them separately and does not choose between these two possibilities.

[8] Although Hunter's argument can be extended either way with justice, he emphasized the influence of demand over the influence of supply. To the extent that the difference between East and West is found to be on the supply side, and the termination of this difference to be the result of changes in supply curves, he may be said to have been misleading. Chapter 2 has verified that the pattern of demand was changing in this period, but this change may have favored the introduction of mineral fuel both in the East and in the West, leaving the lag in the Western reaction to be explained by other means.

in the production of iron with charcoal. This assertion does not follow directly from the argument on quality that Hunter presents, and needs to be demonstrated. It is possible that the factors he cites were not strong enough to control the profitability of using coke. Less than half the pig iron produced in this era was used in the form of wrought iron; in 1849, the amount of pig iron used to make castings was 56 per cent of the pig iron produced.[9] The demand for pig iron was therefore not the same as the demand for wrought iron, and it is necessary to examine the demand for pig iron directly. As Hunter implies, however, it is more plausible to expect quality differences in the iron produced to explain price differentials than it is to expect them to be the result of the tariff. There was no distinction made between various types of pig iron for the purpose of levying duty upon them, and price differentials would be expected to be preserved unless there was a difference in the elasticity of demand for the two types of iron, which in turn would be produced only by prejudice or by a difference in quality. As there was an ascertainable quality differential between coke and charcoal pig iron, we may follow Hunter in attributing the price differential we will observe shortly to it rather than to prejudice. By so doing, we neglect the influence of the tariff, except as it helped produce a high rate of growth in the iron industry, a rate of growth that should have been conducive to technological change.

The Hot Blast

We may now turn from our discussion of previous explanations of the delay in adopting coke to the problem itself. We shall attempt to derive a direct estimate of costs in order to see if relative prices before about 1860 made the production of iron with coke unprofitable. If so, we may ask why this price structure existed and whether it was changes on the supply side or on the demand side that altered the profitability of making coke pig iron in the 1850's. If it was not unprofitable to use coke, we must reexamine our reasoning in the hope of discovering why coke was not adopted before it was. One more preliminary task remains, however, before we can pass on to the consideration of costs and profits; we must understand the

[9] Appendix C, Table C.8.

technology in question before we can understand the structure of costs.

The typical blast furnace of 1830, nearly indistinguishable from its counterpart of fifty years earlier, was built of masonry. It was about 30 feet high, 20 feet along each side of its square base, and it sloped inward slightly as it rose. The interior of the furnace measured only about nine feet across at its widest point (the "bosh") and was lined with sandstone or soapstone, the forerunners of refractory brick. Iron ore, charcoal, and limestone were thrown into the open top of the furnace by men who had carried baskets of these materials across a bridge built from a rise in the ground to the top of the furnace. Air was pumped into the furnace near the bottom by means of an opening (the "tuyere") made of cast iron or brass and not yet protected by water-cooling. The air was supplied by bellows or wooden tubs powered by a water wheel. This blowing apparatus was shortly to be superseded by machines made of cast iron, often employing steam engines for power. The pressure produced by the blowing engines was only about one or one-and-one-half pounds per square inch; stronger pressure was not desired for fear of "blowing the charcoal to pieces."[10]

The first major change in American blast-furnace practice was the introduction of the hot blast, first introduced in England in 1828. This modification of blast-furnace technique was based on a very simple idea. The furnace was operated by blowing a blast of air into a column of ore and fuel in order to produce combustion. The materials had to be heated to burn properly, and part of the energy in the fuel was being used to heat the blast-furnace charge and the incoming air to the point where the desired chemical reactions could take place. If the blast was already heated before it entered the furnace, it would heat the charge and lessen the amount of fuel necessary to smelt the iron.

This innovation could be attached to any existing blast furnace. In its simplest form, it was just a set of pipes over a fire in which the blast was heated on its way to the furnace. A more economical version, because it did not use any extra fuel, was to place the pipes through which the blast passed at the top

[10] Arthur Cecil Bining, *The Pennsylvania Iron Manufacture in the Eighteenth Century* (Harrisburg: Pennsylvania Historical Commission, 1938); *Bulletin, 19* (October 7, 1885), 266.

of the blast furnace where they could be heated by the combustion of the waste gases from the furnace itself. It was in
this form, and more efficient later modifications, that the innovation was generally adopted in America. Once the usefulness
of waste furnace gases was seen, they were employed for making
steam to provide power for a variety of machines, including
the blowing engines.

News of the hot blast was available in the New World with
little delay. Articles appeared on savings in England and on
the necessary machinery, and in 1834 a New Jersey blast furnace adopted the first practical hot blast in the United States.
As modified in the following year to use waste gases to heat
the blast, the innovation saved the furnace 40 per cent on its
fuel costs.[11] But despite the impression created by savings
such as these and by the articles claiming savings up to 75 per
cent for the Clyde Iron Works in Britain, the hot blast was
not a universal benefit. Overman summarized the incidence
of its effects:

> The economical advantages arising from the application of hot
> blast, casting aside those cases in which cold blast will not work
> at all, are immense. The amount of fuel saved, in anthracite and coke
> furnaces varies from thirty to sixty per cent. In addition to this, hot
> blast enables us to obtain nearly twice the quantity of iron within a
> given time than we should realize by cold blast. These advantages
> are far more striking with respect to anthracite coal than in relation
> to coke, or bituminous coal. By using hard charcoal, we can save
> twenty per cent of fuel, and augment the product fifty per cent. From
> soft charcoal we shall derive but little benefit, at least where it is
> necessary to take the quality of iron into consideration.[12]

The primary interest of the hot blast is in those areas where
its effect was the greatest, that is, in connection with mineral
fuel. In fact, before the application of the hot blast to the smelting of iron with anthracite, it was not practical to use that fuel
at all. In such a case, the effects of the hot blast are difficult
to separate from the effects of the use of mineral fuel, and the
two will be treated together. But the hot blast alone is of some
interest.

[11] *JFI, New Series*, 5 (1830), 215; 9 (1832), 339; *10* (1832), 130; *18* (1836), 127;
Swank, p. 453. The full titles of frequently cited periodicals can be found in the
Bibliography.
[12] Overman, p. 442.

The convention of 1849, which collected so much useful information, classified the charcoal furnaces listed by the nature of the blast. From this, two things emerge. There is no difference in the percentage of old and new firms which adopted the hot blast, demonstrating that the hot-blast apparatus could be added to old furnaces without much trouble. But there is a sizable difference in the extent to which the hot blast was adopted in different regions of Pennsylvania. If the state is split into three—the East, the Juniata Valley, a region in the mountains which produced high-quality iron for sale in both Philadelphia and Pittsburgh, and the remainder of the West— we find that in the first two regions over half of the charcoal furnaces used the hot blast in 1849, but that in the last region only one-fifth did so.[13] We may ask why this differential exists.

The cost of adopting the innovation was not high, as was argued above and as can be seen from a direct examination of the convention's data. The increased capital costs were less than 10 per cent of total capital costs, and were more than compensated for by the increased output resulting from the hot blast.[14] The extant statistics are not complete enough to allow a detailed cost calculation in light of the variable nature of this innovation, but we may isolate two possibilities why the hot blast was not used in the West as much as in the East, paralleling the two possibilities that will emerge from our discussion of coke. The raw materials in the West may not have been suitable for the hot blast, that is, there may have been the wrong kind of wood to make charcoal, or there may have been a difference in demand that deterred the use of this innovation in the West.

Quality differences existed between some hot-blast and cold-blast iron, and continuing discussion of quality problems supports the contention these differences persisted over time.[15] The extent to which they were reflected in prices, however, remains obscure. When the process is specified, which is far

[13] Convention of Iron Masters, *Documents Relating to the Manufacture of Iron, Published on Behalf of the Convention of Iron Masters which Met in Philadelphia on the 20th of December, 1849* (Philadelphia, 1850), Tables. The differences among regions were statistically significant at the 5 per cent level, while intertemporal differences were not.

[14] *Ibid.*

[15] *JFI*, N. S., *24* (1839), 334-345; Overman, for example, pp. 436-440; *Iron Age* (March 11, 1875), p. 23.

from universal, only a slight differential is detectable.[16] In addition, as the passage from Overman suggests, this quality differential was related to the quality of the charcoal used. It was also, not surprisingly, related to the quality of the iron ore used. These differences are all sufficiently small that, at this level of investigation, it is not possible to separate their effects and to isolate differences in demand or differences in raw materials as the cause of the observed inter-regional difference. While such a situation is lamentable, the hot blast derived its major importance from its permissive effect on the use of mineral fuel, and we hope to have better luck in our investigation of the larger question.

In the course of the 1830's a Welshman and an American discovered independently that the hot blast enabled anthracite to be used in the blast furnace. Although coke was well-known as a blast-furnace fuel by then, the difficulty of igniting anthracite had prevented its similar use. The American inventor died before he could test the commercial feasibility of his invention, but the Welsh inventor, David Thomas, was brought to the United States, where he built the first anthracite furnace that was both a commercial and a technical success in 1839. Its distinguishing features were its large size and powerful blast: the latter was six pounds per square inch, according to one report. "With the erection of this furnace commenced the era of higher and larger furnaces and better blast machinery, with consequent improvements in yield and quality of iron produced."[17]

The construction of this furnace, and specifically its important blowing engines, was quite difficult. Much of the machinery was brought from England, and arranging to have the rest built in this country posed many problems. The cylinders for the blowing engine are a case in point. Thomas wanted cast-iron cylinders of five-foot diameter and six-foot stroke. He arranged to have them brought from England before he left for the United States, but they were too large to go through the hatches of the ship hired and were not brought over immediately. Thomas tried to have the cylinders made in this country, but there were

[16] Louis C. Hunter, "A Study of the Iron Industry of Pittsburgh Before 1860," unpublished doctoral dissertation, Harvard University, 1928, pp. 392-433.

[17] Swank, pp. 354-361; William Firmstone, "Sketch of Early Anthracite Furnaces," *AIME*, 3 (October, 1874), 152-156, esp. 155; *Iron Age*, 43 (March 7, 1889), 348.

no boring-mills in the United States that were big enough for the job. The Southwick foundry of Philadelphia finally undertook the job, enlarging their boring-mill for the purpose. The price of the cylinders was $.12½ a pound, or $280 a ton.[18]

After these difficulties and his famous commercial success, Thomas continued his leadership of the anthracite region, blowing in a total of five furnaces for his employers and forming the Thomas Iron Company in 1854 with his sons.[19] His example was widely followed and the amount of iron made with anthracite rose rapidly. The contrast of this movement with the continuing neglect of the new technology in the West is striking and should be reflected in the price and cost structures of the two regions.

Costs and Profits: A Calculation

The first question to be answered is whether or not it was profitable to use coke during the great expansion of the iron industry in the 1840's. The use of anthracite in this period implies that it was profitable, and we may use this inference as a check on our results. Each of the new processes, however, must be compared separately with charcoal, not with the other new process. The two types of coal were found on opposite sides of the Allegheny Mountains and it was not feasible before the middle 1850's to transport large amounts of coal or iron across the mountains. Charcoal was produced on both sides of the mountains and used for the production of iron in both regions; it was the competing fuel for both types of coal. As time went on and transportation improved, the two types of mineral coal would come to compete with each other.

Not surprisingly, direct profit figures are not available. It is therefore necessary to reconstruct them. I propose to do this by subtracting the costs of production from the selling price, getting the profit margin per ton of pig iron produced. Then, by noting any change in the amount of pig iron produced per unit of capital, it will be possible to compile an index of the difference in profitability between the various iron-produc-

---

[18] Samuel Thomas, "Reminiscences of the Early Anthracite-Iron Industry," *AIME*, 29 (September, 1899), 901-928. The normal price for castings was about $70 a ton, or one-fourth the price of the cylinders. (Berry, p. 276.)

[19] Swank, pp. 360-361; *Bulletin, 16* (June 28, 1882), 173.

ing techniques. The last step will be found to be an important one, as output per unit of capital changed quite drastically, and it is not important if the cost calculations are, as they are, only approximate.

As Hunter noted, the quality of iron produced with different fuels differed, and this difference was reflected in the market price, the greater impurity of coke, or the sensitivity of demand to quality in the West, yielding a lower price for coke pig iron than for anthracite pig iron.

The prices come from two sources. From the American Iron and Steel Association come Philadelphia prices for charcoal and anthracite pig iron that overlap for six years in the 1840's. And from Hunter come newspaper quotations giving prices of pig iron at Pittsburgh in the 1850's. Using those years in which there are price quotations for pig iron made with coke or bituminous coal gives eight years in which there are comparable prices for charcoal pig iron and coke pig iron. Despite large cyclical movements, the price of charcoal pig iron averages out to be about $30 a ton in both markets in both decades, reflecting similar cost structures in the two regions. Anthracite pig iron averaged $28 a ton for the years in which we have overlapping quotations, and coke pig iron averaged only $24. The $2 and $6 differentials are roughly conserved in the year-to-year movements.[20] The average prices have been entered at the top of Table 3.1.

These are prices at the cities; to get the price at the furnace, transport costs must be deducted. A contemporary writer gives the average cost of transporting pig iron made with charcoal and with anthracite to Pittsburgh in 1857 as $5 and the cost of transporting pig iron made with coke to the same city at $1.[21] This is virtually the only information available on the cost of transporting pig iron from the furnace to the market, although one of the blast furnaces known to have been using coke before 1850 was having financial difficulties that were attributed to difficulties of transportation.[22] As there are many diverse elements in the cost of making pig iron and getting it to the market, the attribution of difficulties to the high cost of transport may

[20] Appendix C, Table C.15; Hunter, "A Study of the Iron Industry . . .," pp. 393-433.

[21] George H. Thurston, *Pittsburgh as It Is* (Pittsburgh, 1857), p. 105.

[22] The Lonaconing furnace is meant. Walter P. Johnson, *Notes on the Use of Anthracite* (Boston, 1841), pp. 8-9.

TABLE 3.1

CALCULATED PROFITS FROM PRODUCING PIG IRON, ABOUT 1850

	Type of Production		
	With Charcoal	With Anthracite	With Coke
Selling price per ton	$30	$28	$24
Less: Transport charges	5	5	5
Furnace price per ton	25	23	19
Less: Ore and flux	6	6	6
Fuel	9	7	3
Labor	2	2	2
Interest on fixed capital	2	1	1
General expenses	2	2	2
	21	18	14
Profit margin per ton	4	5	5
Profit rate (Index)	100	250	250

Source: See text.

not be entirely accurate, but on the surface it appears to conflict with a vastly lower cost of transport for furnaces using coke as a fuel.

The lack of evidence about transportation costs indicates that the best procedure is to ignore them and to assume that the price differential in the cities reflects the price differential at the furnaces. This would be accurate if most of the iron prices quoted in the cities were f.o.b. at the furnace, which they probably were not at this time although they were later, or if transport costs were the same for all classes of furnaces, which is precisely what we do not know. Charcoal furnaces probably had higher transport costs than mineral fuel furnaces because they had to be located in outlying regions to have an adequate supply of their nontransportable fuel. This is consistent with the costs quoted above, as the cost for anthracite pig iron was the cost to bring it to Pittsburgh which was presumably above the cost to bring it to Philadelphia, but it conflicts with the other evidence quoted. The procedure adopted here, then, overstates the relative profitability of using charcoal in all probability.

The differentials at the furnace are thus to be considered the same as the differentials in the cities. I have, however, put in a constant transportation charge of $5 a ton to enable the residual obtained to be labeled "profit." Subtracting this charge from the market price gives a figure from which the actual costs of production are to be subtracted.

The requirements for ore and fluxing materials were not significantly altered by the change in fuel or the use of a hot blast; their cost may be safely taken at $6 per ton of iron produced. This figure is an average of various figures, estimated by inspection rather than any rigorous means. It represents the cost of 2 to 2½ tons of ore plus a small charge for limestone or some other flux.[23]

The fuel requirement is the item that varies most between the various techniques. The charcoal input may be taken as 180 bushels, an average estimate from Overman's manual, priced at $.05 a bushel, an average price quoted in this manual and also derived as an average of the charcoal prices reported to a Congressional inquiry.[24] The anthracite requirement was probably under $7, the cost of 2 tons at $3.50 apiece.[25] The amount of coke required to produce a ton of iron cannot be estimated simply. The only direct evidence for this input is that cited by Hunter: three advertisements giving costs ranging from about $2 to $5.[26]

Hunter cites some English prices, but these clearly are not relevant to a consideration of profitability in western Pennsylvania. The $5 fuel cost comes from Missouri; it, too, is from a distant source and therefore only relevant to a small part of the problem at hand. The remaining two estimates are the important ones. The lowest one comes from the Lycoming Iron and Coal Company, of Lycoming County, Pennsylvania. The cost is $.60 a ton for coal, 3½ tons of which were needed to smelt a ton of iron, making the cost $2.10. The Lycoming Company, however, is not listed as operating a coke blast furnace in either of the standard sources for this period. The 1849 con-

---

[23] See, for example, B. F. French, *Rise and Progress of the Iron Trade of the United States from 1621 to 1857* (New York, 1858), pp. 63, 100; Allan Nevins, *Abram S. Hewitt* (New York: Harper and Brothers, 1935), p. 101.

[24] Overman, pp. 117, 151; U.S., Congress, House, Executive Document No. 2, 36th Cong., 2nd Sess. (1860), pp. 130-153.

[25] French, pp. 63, 100; Nevins, p. 101; Johnson, pp. 10-11; Samuel Harries Daddow and Benjamin Bannan, *Coal, Iron, and Oil, or the Practical American Miner* (Pottsville, Pennsylvania, 1866), p. 715.

[26] Hunter, "The Influence of the Market . . .," p. 263.

vention recorded that they owned a charcoal furnace at Ralston, the site of a bituminous coal deposit. Lesley reports the Ralston Furnace as a charcoal and anthracite furnace and observes that it was abandoned. The furnace Lesley referred to was one built to replace an older furnace which burned down in 1853-1854, and it used anthracite for fuel. "After two unsuccessful blasts the furnace was abandoned. The ore is a carbonate of iron in solid layers and in balls, underlying the bottom conglomerate of the coal measures."[27] The failure of the furnace and the fact that it used anthracite for fuel cast doubt on the availability of a bituminous coal suitable for blast furnace use at the cited price.

The remaining cost citation comes from the Cambria Iron Works at Johnstown, Pennsylvania. They used 5 tons of coke per ton of pig iron produced at $.54 a ton, yielding a cost of $2.70 per ton of iron. Cambria was not without its financial troubles, but after it started using coke in 1853-1854 it rapidly expanded to become one of the largest ironworks in the country.[28] This cost is then a reliable one, but may not be representative. On the other hand, other ironmasters were free to locate where they thought the best raw materials could be obtained, and there is no indication that the Johnstown coal was the best available. The major coal mines of the post-Civil War era were located at some distance away.

Let us follow Hunter and use $3 as the cost of coke. This will give us a provisional cost, although not necessarily one that was attainable everywhere. The meaning of this "attainable cost" will become clearer when we return to this point later in the discussion.

The cost of labor can be taken to have been $2 per ton for all three processes.[29] There is every reason to expect that labor costs were lower for furnaces using mineral fuel as their output was larger, while the use of labor per blast furnace was relatively stable. But in the absence of any evidence on wages to

[27] Convention of Iron Masters, *Documents* . . ., Tables; J. P. Lesley, *The Iron Manufacturers Guide to the Furnaces, Forges and Rolling Mills of the United States* (New York, 1859), p. 93. The Ralston coal deposit is described in Daddow and Bannan, p. 312.

[28] The history of this firm is discussed more fully in Chapter 5.

[29] French, pp. 63, 100; W. M. Grosvenor, *Does Protection Protect?* (New York, 1871), p. 233; Overman, pp. 151, 182.

this effect, the consideration is ignored, despite the fact that output per worker was more than doubled by the change in fuels.[30]

Interest on fixed capital was deduced by applying a fixed interest rate, of 6 or 7 per cent, to the investment in land and buildings per ton of productive capacity as shown by the 1849 convention of ironmasters. As can be seen from the table, the output per dollar of investment was doubled by the switch from charcoal to mineral fuel. As this point is a critical one in the argument, a careful consideration of it will be postponed until its importance can be demonstrated.

General expenses include everything that has been left out, except depreciation, which was not usually calculated at the time and is hard to reconstruct now. (It is apparent that in view of the small cost of capital the influence of this factor would be quite limited; in addition, it would increase the cost of using charcoal due to the greater amount of capital required per ton of output.) General expenses thus include such items as inventory shrinkages and losses, interest on inventories of raw materials, interest on the stock of pig iron, interest on sales for which long credit was extended, reserve for bad debts, possible brokerage fees, administrative expenses, etc. This was usually given as $2, and it is entered in Table 3.1 as such.[31]

Performing the appropriate subtractions, the profit margin per ton can be calculated as shown in Table 3.1. The important point is that they were not lower for production with mineral fuel than with charcoal. This result is supported by some additional evidence pointing toward a higher cost for furnaces using charcoal in the region where coke was used. Charcoal consumption was not uniform across geographical areas, the West using more than the East.[32] This was possibly related to the less extensive use of the hot blast west of the Alleghenies than in the East, but we do not know. Capital conditions were also different on different sides of the mountains, capital being

[30] Convention of Iron Masters, *Documents* . . ., Tables.

[31] French, pp. 63, 100; Nevins, p. 101. The $4 figure quoted by Nevins includes both general expense and labor. The list of items included in general expenses is adapted from a comment on later cost calculations: *Iron Age, 42* (November 1, 1888), 669.

[32] Overman, p. 151; A. J. Davis (editor), *History of Clarion County, Pa.* (Syracuse, 1887), p. 113.

scarcer in the West. To the extent that capital costs were higher, the difference in capital costs was larger.[33]

It is necessary to convert the profit margins shown into profit rates on invested capital, taken to be fixed capital because of the lack of any other kind of data. As will be seen shortly, the dollar value of fixed capital required per ton of output was only half as great for furnaces using mineral coal as for furnaces using charcoal. (Other capital requirements can perhaps be assumed to move in a similar direction, due to the larger scale and reduced seasonality of production with mineral fuel.) Similar profit margins per ton of output then meant a profit rate for blast furnaces using anthracite or coke that was twice as large as the rate for those furnaces using charcoal. This is a big difference, a powerful inducement for technological change.

The capacity figures for the furnaces using charcoal and anthracite are taken from the 1849 convention; the figures for the furnaces using coke have to be derived from other sources. The convention reported the investment in land and buildings for 295 blast furnaces in Pennsylvania, of which 150 were in blast in 1849, compared with 180 furnaces in that state reported to be in blast by the Census of the same year.[34] Due to the location of the coal deposits, we are concerned primarily with Pennsylvania, as it was in that state that the transition away from charcoal was to be made. As a bonus, this sample represents about half of the iron industry in the whole country.

Investment figures were obtained for 230 charcoal-using blast furnaces; the average investment in land and buildings was about $38,000. The comparable figure for the furnaces using anthracite was over $55,000, or almost half again as much as the investment per charcoal-using blast furnace. But the capacity of the furnaces was increased more than their cost in the change from one process to the other; while blast furnaces using charcoal could produce an average of 1,350 tons of pig iron a year apiece, blast furnaces using anthracite could make an average of about 3,800 tons, or almost three times as much. This means that the investment per ton of capacity was about $30 in a

---

[33] Unless, of course, the supply of capital was more inelastic. See Hunter, "Financial Problems of the Early Pittsburgh Iron Manufacturers," *Journal of Economic and Business History*, 2 (May, 1930), 520-544.

[34] U.S. Census, Seventh, *Compendium* (Washington, 1854), p. 181.

charcoal-using blast furnace, but only about $15 in a furnace using anthracite.[35]

These figures are subject to the inevitable criticisms: they represent only fixed capital, and possibly not even the entirety of that if tools and machines were important; we are not told if the investment is evaluated at original cost, depreciated value, or replacement cost; we do not know how they were compiled; and we have no independent test of their accuracy. Nevertheless, they represent more than could really be hoped for under the circumstances. We can be reasonably sure they came from the ironmasters themselves, both from the form of presentation and the character and completeness of the data. And the wide coverage indicates that the conclusions drawn from these figures can be applied to the industry (of Pennsylvania) as a whole.

Unfortunately, the convention reported only eleven blast furnaces using coke or bituminous coal and apparently made an error in reporting the four coke furnaces.[36] This is, of course, a reflection of the fact that anthracite had been widely adopted as a blast-furnace fuel by 1849, while bituminous coal and coke had not. It is necessary to examine those furnaces using coke about which information has been preserved to see whether they were more similar to blast furnaces using anthracite or to blast furnaces using charcoal.

To do this we must convert the information reported by the 1849 convention on annual production to data on weekly production. This involves information about the proportion of the year for which furnaces were "in blast." If the annual capacity figures cited above are divided by 52, minimum weekly capacity

[35] Convention of Iron Masters, *Documents . . .*, Tables. In this calculation I have not separated regions or charcoal furnaces using hot and cold blasts. In the East, investments per tons of capacity were $28 and $34 for the two types of charcoal furnaces. In the western part of the state, charcoal furnaces were were slightly less expensive than in the East and represented $21 or $27 investment per ton of capacity, the latter referring to the cold-blast furnaces that formed the overwhelming majority of charcoal-using furnaces in this region. The investments per ton of 1849 production were $30, $62, and $80 for anthracite, hot-blast charcoal, and cold-blast charcoal blast furnaces.

[36] Convention of Iron Masters, *Documents . . .*, Tables. The furnaces using bituminous coal have smaller investments per ton of capacity than do the charcoal-using furnaces, as is to be expected. But the furnaces using coke, all operated by the Brady's Bend Iron Company, have much larger investments. This unexpected result appears to come from including part of the Brady's Bend rolling mill in with the blast-furnace investment.

figures compatible with them are obtained. They are about 70 tons a week for furnaces using anthracite, and about 25 tons a week for furnaces using charcoal as fuel. Overman states that the capacity of blast furnaces using anthracite, and also those using coke, was 70 to 80 tons a week.[37] This suggests both that mineral coal furnaces were in blast most of the year, and that coke-using blast furnaces had capacities similar to anthracite-using furnaces. The task is now to confirm this capacity estimate for coke furnaces, showing that the annual capacity was at least equal to the annual capacity of furnaces using anthracite. (The actual average weekly capacity for charcoal-using blast furnaces is usually given as 25 tons, but there was a period each year when the furnace was shut down and charcoal was made. The capacity cited above for charcoal-using furnaces is thus an overestimate.)

Of the companies with blast furnaces using coke before 1850, information has been preserved for only a few. Several of these experimented with coke, but did not make a serious attempt to use coke for an extended period of time. William Firmstone is said to have made "good gray forge iron" with coke for a month in 1835, and F. H. Oliphant to have made about 100 tons of coke pig iron about 1837. Other ironmasters conducted similar experiments, primarily in southwestern or north-central Pennsylvania, but their furnaces, like the anthracite furnaces built before Thomas' famous furnace, were not commercially successful even when they were technical successes. One of the most spectacular failures occurred at Farrandsville, not too far from the location of the Lycoming Iron and Coal Company's furnace, where a large furnace was built in 1834 "at great expense, with apparatus and surroundings on the same scale, in confident expectation of smelting coal measure ores with semi-bituminous coal or coke, and at least half a million dollars were expended before proper experiments had tested the quality of the coal and ore beds. In the end the whole was abandoned about 1838."[38]

The first commercial success in the use of coke in the blast furnace was achieved at the Lonaconing furnace in western Maryland. This furnace played the same role in the use of coke that Thomas' furnace played in the use of anthracite. It was

---

[37] Overman, pp. 177, 179.

[38] Swank, pp. 366-369; Lesley, p. 93. The Farrandsville furnace is listed as a charcoal furnace that was originally built as a coke furnace by the 1849 Convention of Iron Masters.

the first one to be a large-scale furnace with heavy machinery and to make the introduction of mineral fuel the occasion for the simultaneous introduction of several other scale-enlarging innovations. The Lonaconing furnace was built in 1837, slightly earlier than Thomas' furnace, was 50 feet high and had a blast heated to a temperature of 700 degrees fahrenheit and powered by an engine of 60 horsepower. In 1839, it was making 70 tons a week of good foundry iron, the quantity that was cited above.[39] Two other firms used coke in their blast furnaces before 1850; Overman believed their furnaces to be copies of the Lonaconing furnace. They had almost the same dimensions and their production was also about 70 tons a week.[40] The furnaces using coke in the 1840's were thus very similar to the furnaces using anthracite with respect to scale.

We may make the assumption that the cost of a coke furnace relative to a charcoal furnace in the same region was the same as for anthracite furnaces, as indicated by the similarity in operating characteristics of the two kinds of mineral coal furnaces. This extension allows us to say that a margin of profit per ton of iron made with mineral fuel meant twice as much as an equivalent margin of profit per ton for charcoal pig iron. An index of profit rates may then be calculated from the profit margins per ton shown in Table 3.1. It shows that it was possible to greatly increase the return to capital by using either anthracite or coke to make pig iron instead of charcoal. These calculations were performed for large-scale mineral fuel furnaces, and they are therefore valid only for the time after the discovery that this style of furnace was needed in the late 1830's.

Costs and Profits: Further Considerations

The argument has not answered the question that was asked; instead, it has posed a paradox. Presented with two situations, the reactions to which were strikingly different, the argument using fuel costs suggested by Hunter has failed to discover a difference between them. By changing the fuel used from charcoal to either type of mineral fuel, the pre-Civil War ironmaster was able to greatly increase his output. This represented an increase in the return to his capital because his profit margin

[39] Swank, pp. 369-370; Johnson, p. 7.
[40] Overman, pp. 175-177.

per ton of iron produced did not decline, the price decline in each case being balanced by a decline in fuel costs of roughly similar magnitude.

The increased profit was dependent upon the increase in scale for several reasons. The leverage of the profit margin per ton of iron made was increased, and the profit margin itself was increased. The interest on fixed capital per ton of iron produced was larger for firms with smaller outputs, and it may be presumed also that the smaller furnaces had less powerful and cooler blasts and therefore used more fuel. We would expect, then, that the use of mineral fuel was not profitable unless combined with other innovations that increased the scale of operations. This, in fact, was true. The anthracite furnaces built before Thomas' furnace were built on a scale similar to charcoal furnaces, made under 40 tons of iron a week, and were unprofitable.[41] We do not know the technical characteristics of the coke furnaces built before the Lonaconing furnace, but the attention given to the scale of this furnace indicates that they were very similar to charcoal furnaces. The cost calculations in Table 3.1 are thus able to answer why coke was not adopted before about 1840, even if they cannot answer why it was not adopted at that time. The relative price structure in the United States was different from that in England, and it was not profitable to simply substitute mineral fuel for charcoal. Only after the introduction of several other innovations, such as the hot blast and powerful blowing engines, was it profitable to use mineral coal in the United States.[42]

But the paradox remains: if it was profitable to use both types of mineral fuel, why was only one adopted in the great expansion of iron production, in both East and West, in the 1840's? In the discussion of the cost of coke, the costs derived were referred to as "attainable costs." The obvious inference from the contrast between the attitude of ironmasters to anthracite and coke is that these costs, or at least the profit shown, were attainable by only a very few people in the West. An argument that makes the ironmasters of this time studiously ignore a profit differential of the size shown is not reasonable in the light of all else we know about the American economy.

[41] Johnson describes some of these furnaces. Firmstone uses Johnson's data. Swank, pp. 358-360, summarizes Johnson and Firmstone.

[42] Overman, p. 401, talks about the need for a powerful blast. He sees it as a technological necessity, rather than an economic one.

A rapidly expanding economy containing entrepreneurs who are seeking economic advantage may be expected to respond rather quickly to changes in the price and profit structure of the economy. It is in backward economies, sectors, or industries—where we find the parts of economies that are slowly being eliminated by the force of competition—that we find people paying little attention to market forces. If the model of competition has any usefulness, it must imply that those people, industries, and economies that succeed in the marketplace are acting in accordance with economic incentives. We know that the inhabitants of the West were alert to all manner of change in this period, even in other branches of the iron industry itself, and it is unreasonable to expect a pocket of irrationality or ignorance to have remained in this restricted sector.

There is no evidence of a large gain in profitability for those furnaces which actually used coke in this period. The Lonaconing furnace, which made a good foundry iron, found that the high cost of transport had "paralyzed its operations" in 1841.[43] The furnaces at Mount Savage and at Brady's Bend were not able to produce a grade of foundry iron salable at a remunerative price, due to the low quality of their coal, but were able to make iron that they could use in their own rolling mills. Overman comments that "it is a riddle among iron manufacturers, why these establishments were erected at the precise place where these natural difficulties [inferior coal] can never be removed."[44]

The financial histories of these two firms are confused and their difficulties may be due partly to the troubles of operating the two first rail mills in the country. To the extent, however, that these firms made rails because of a prior decision to use coke in their blast furnaces, this is a distinction of little importance: the risks of making rails were part of the risk of using coke. Mount Savage failed in 1847 and was out of operation in 1849; Brady's Bend was also out of blast in 1849 and had some difficulties signalized by the confusion in its name which appears to have changed from the Great Western Iron Works to the Brady's Bend Iron Works around 1847. Both firms were owned by Boston capitalists by the end of the 1840's: Mount Savage by J. M. Forbes, and Brady's Bend by M. P. Sawyer.[45]

[43] Johnson, pp. 8-9.
[44] Overman, p. 175.
[45] *Hunt's, 16* (1847), 212; *21* (1849), 460-461; *ARJ, 20* (1847), 737; Lesley, p. 96; Hunter, "A Study of the Iron Industry . . .," p. 113.

They both, despite their financial difficulties, remained in existence and became large rolling mills. These firms were the most successful coke users of the 1840's.

The figures derived above on profits were thus attainable by only a few people, if at all; the demand or supply curves, possibly both, must have been very inelastic at the prices and costs shown. The high calculated profitability of using coke depended upon three factors: a multiplication of output by a large amount, a price differential of $6, and a cost of coke of $3. The first of these was achieved at the furnaces in question and was the result of a technique we may assume was known widely after 1840. The second is an average figure, with all of the attributes of averages, but is larger than the $5 (or less) differential usually assumed in contemporary discussion.[46] In only one of the eight years for which Hunter has comparable prices for coke and charcoal pig iron was the differential more than $7.[47] And a difference of $1 is not large enough to reverse the conclusions reached in Table 3.1.

The cost of coke, however, was based on very restricted evidence, as was noted when the data were presented, and the difficulties of the Mount Savage and Brady's Bend works suggest that good coke was not generally available at a suitable price. Overman was unable to decide why these firms had located where they did, but it is not unreasonable to infer that they did not know of any better locations. The centralization of coke production for blast-furnace use in the celebrated Connellsville region after the Civil War was the result of the superiority of that area's coke over the coke produced from other locations, but Connellsville coke was not discovered until the early 1840's, and it did not immediately find general favor.[48] Geographical exploration is not accomplished in a day, and if this variety of coke was sold for the first time only in 1842, it should be no surprise that the boom of the middle 1840's was accomplished without its help.

Other coke can be used in the blast furnace, but if the coal used is inferior, special care—which implies extra cost—must be taken in the coking process, either to assure a suitable physical structure in the coke produced or to remove sulphur from

[46] For example, Lesley, pp. 759-760.
[47] Hunter, "A Study of the Iron Industry . . .," pp. 393-433.
[48] Swank, pp. 476-477.

the coal. The late entry of the Connellsville coke region on the scene may have been the result of inadequate geographical knowledge, but it also may have been the result of an inadequate appreciation of the difficulty of coking inadequate coal. This area of indeterminacy cannot be resolved with the data at hand, but in the light of the success enjoyed by other branches of the iron industry in adopting British innovations and the use of coke for other purposes in America, it seems likely that the knowledge lacking involved an appreciation of the natural resources of the West rather than an understanding of the coking process itself.

The importance of the extra costs from inadequate coal, however, is problematical. Any difficulties arising from the physical structure of the resultant coke should have been included in the costs of production reported above. And quality loss due to an impure fuel was allowed for in the $6 price differential for the iron produced. We must conclude either that the coke available at the price shown could not be used to produce iron at the costs shown or that coke was not generally available at this price.

The latter choice appears to be the relevant one. Overman, for example, reported that the price of coal at the coal pits near Pittsburgh was $.50 to $1 per ton, one-quarter to one-half as cheap as anthracite coal. Fifty per cent of the weight was lost in coking, and the cost of mining coal for one ton of coke was therefore $1 to $2.[49] This cost is twice the cost of coke quoted by Hunter, and it does not even include the cost of coking itself. If Overman's costs are typical, they indicate that the cost cited by Hunter reflected a special situation, and that the error in the profit calculation above was in the cost of coke. More correctly, they indicate that the cost derived above was attainable by only a few people in fortunate locations and that the supply curve at the cost shown was therefore very inelastic. (The effects of the introduction of Connellsville coal in this situation will emerge shortly.)

To identify the source of confusion in the data, however, does not resolve the historical question. For coke was as cheap or cheaper than anthracite, and the unprofitability of its use cannot be attributed solely to its cost. The principal difference between East and West, using revised coke costs, was the greater

[49] Overman, pp. 102, 130-132.

price differential between different types of iron in the West. We must ask whether this price differential was due to the lower quality of the mineral coal available in the West or to a difference in the pattern of demand.

Fortunately, a direct answer can be made to this question. Coke pig iron was never sold in the East before the Civil War, but pig iron made with anthracite was imported to Pittsburgh in the 1850's. This type of iron was sold on a continuing basis in Pittsburgh after 1853, and it sold in that city for the same price as charcoal pig iron, rather than at the price of coke pig iron.[50] This indicates that the purchasers of pig iron in the West considered anthracite pig iron to be of the same quality as charcoal pig iron and of substantially better quality than pig iron made with coke. In other words, had pig iron made with coke been of the same quality as pig iron made with anthracite, buyers in the West would have paid substantially higher prices for it, and the profits from making it would have been correspondingly greater.

The character of demand in the two regions may have been different—indeed, it probably was—but the greater price differential between charcoal and mineral fuel pig iron in the West must be attributed to the lower quality of the mineral fuel pig iron made with the coal of that region, that is, to differences in the supply curves of the two regions. We have emphasized in the foregoing discussion the great impurity of the available bituminous coals, which contrasted sharply with the relative purity of the anthracite coal of eastern Pennsylvania. This was the cause of the low quality of coke pig iron. The contrast between the regions in this period, then, was due primarily to a difference in their known resources, not a difference in the knowledge of production techniques or the character of demand. The backwardness of the West may even be attributed to the progressiveness of the East. For the new technology of the age was not being neglected, it was being applied where it was most easily used. If it had not been possible to expand the production of iron with anthracite in this period, the resultant bottleneck might have stimulated people to find new sources of coal and to use bituminous coal earlier than they did.

[50] Hunter, "A Study of the Iron Industry . . .," pp. 393-433.

The Adoption of Coke

This, of course, raises the final question of fuel to be considered. By the close of the Civil War, coke was beginning to be used as a blast-furnace fuel with rapidly increasing volume; some factor must have changed over the course of the 1850's to produce this change in attitude toward coke. A difference in raw material availability does not necessarily mean that that imbalance has to be corrected for the differences in production to be eliminated. An offsetting factor may easily arise. Three factors appear as likely candidates to change the relevant expectations: increased competition from the East, discovery of better coke, and a rise in demand for low-cost pig iron. All three appear to have been present.

The railroad from Philadelphia reached Pittsburgh in 1852, and from that time on anthracite iron was quoted in the Pittsburgh market. The resulting downward drift of prices supplanted the effects of British competition and helped produce the rapid decline in the production of iron with charcoal that was noticed in Chapter 1. The railroad had a dual effect: it increased the demand for iron products in the West by stimulating the economy, but it reduced the demand for iron products relevant to Western ironmasters by making Eastern supplies more accessible to Western markets. The Western ironmasters were forced to change their production techniques and lower their costs or die, as it were, in the midst of plenty. The decline in the production of charcoal pig iron by one-half between the peak rate in the 1840's and the outbreak of the Civil War was one side of this coin.[51] The extension of coke production at Connellsville that was such a notable feature of the post-Civil War iron industry was the other.

It was mentioned above that the use of inferior coke led either to higher costs of manufacture or to lower quality iron and consequently lower prices for the iron produced. The exploitation of the Connellsville coal fields did not, so far as is known, alter the costs of making pig iron with coke. Instead, it raised the quality of the pig iron and the price for which it sold. Prices are not available for the Civil War or for several years thereafter. They become available in the course of the

[51] See Chapter 1 and Appendix A.

1870's, and at that time they exhibit a markedly different pattern than the one they showed in the ante bellum years.

Before the Civil War, charcoal pig iron was the most expensive of the kinds of iron made, with anthracite pig iron selling at a discount of under 10 per cent (in the East) and coke pig iron selling at a discount of about 20 per cent. When anthracite pig iron was sold in the West, moreover, it sold for the same price as charcoal iron. In the market reports of 1874, the first available after the war, pig iron made with coke sold for the same price as pig iron made with anthracite, both selling at a discount of about 20 per cent or more from the price of pig iron made with charcoal. The prices of anthracite pig iron and coke (or bituminous coal) pig iron initially were reported separately, but they were quickly amalgamated into a single category.[52]

Looking at the supply side first, the effect of having the same price for pig iron made with anthracite and pig iron made with coke was to encourage the production of the latter in the West. As the two prices approached each other in the years between 1850 and 1870, the use of coke in the blast furnace became relatively more profitable until it surpassed the level needed to induce people to use it. The ease of transporting iron over the Allegheny Mountains, however, discouraged the production of pig iron with charcoal in the West before the changing price structure led to an exploitation of the coal resources of that region, the former happening in the 1850's and the latter after the Civil War.

Turning to the demand side, we may ask why people were willing to pay the same price for pig iron made with the two mineral fuels after the Civil War when they had not been willing to do so before. The answer must be that the quality differential between them had disappeared. It may be misleading to attribute the change in the quality of coke pig iron entirely to the use of Connellsville coke, but in the absence of other identifiable factors, this one must be weighted heavily. The improved quality of coke pig iron, therefore, stemming at least partly

[52] The first regular market report appeared in *E&MJ*, 17 (1874), 55. It assumed its complete form a month later, in *E&MJ*, 17 (1874), 134-135, when the two types of mineral fuel pig iron were classified together in the Pittsburgh report. An earlier report of sales appears in *Iron Age*, 11 (1873), 19. It shows the prices of the two types of mineral fuel pig iron at about the same level, but does not include the price of charcoal pig iron.

from the introduction of Connellsville coke, was an important factor leading to the spread of coke as a blast-furnace fuel.

It is always difficult to say why a new source of raw materials was discovered or exploited. But if an inducement was present, it is surely preferable to regard the extension of an economy's raw material base as induced rather than autonomous. For the development of the Western iron industry, Connellsville coke had to be produced. But the stimulus to this development was growth of transportation and communication with the East, rather than an autonomous pull from the Western coal fields.

In this light, the growing demand for railroad iron in the West may be understood as a concomitant effect of the westward drift of the economy. The production of rails in the West grew at the same time as the production of coke pig iron, and it is impossible to say which led. But it is not important to know whether the discovery of coke stimulated the production of rails or whether the demand for rails stimulated the discovery of coke. Both were effects of the economy's increasing extent, motivated perhaps by agricultural expansion, and their relationship was therefore complex and interdependent. The growth of Western rail mills is described in Chapter 5, but rails could have been supplied from the East. The use of coke as a blast-furnace fuel was not only stimulated by the production of rails in the West, it made this production possible.

We may summarize our conclusions as follows: The reason why coke was not adopted in the 1840's in contrast to the adoption of anthracite was because the iron made with anthracite was of substantially better quality than the iron made with coke, and the price of coke was not sufficiently low to offset this factor. The barrier to the spread of mineral fuel technology to the West was not the different character of demand in the West; it was the different supply conditions in the West, as determined by the technological nature of the available raw materials. The lack of raw materials was remedied in the 1850's and later years by the exploitation of the Connellsville coke region, leading to the elimination of the price differential between anthracite and coke pig iron. The new price structure encouraged the use of coke, and its production expanded rapidly in the years following the Civil War.

It would be unfair, however, to say that industry was pulled into the West to exploit the Connellsville coal fields; their discovery was instead induced by the westward growth of the economy. The main symbol of this growth, as well as an important part of it, was the extension of through railroad lines across the Allegheny Mountains. The extension of the railroads created a demand for rails in the West, and this demand was filled partly from the growing production of coke pig iron. These two, the growth of Western rail production and coke pig iron production, aided each other, but neither was the initial cause of the other. Both were the results of the continuing expansion of the American economy, although it must be noted that both then contributed to the further expansion of this economy.

4

The Independent Blast Furnace

Size

The previous three chapters have described what the iron industry was making and how these products were made insofar as this information helped to explain what was made. This chapter and its successor continue the exploration of how the iron industry functioned, but now in relation to the organization of the industry itself. In terms of the three variables listed in the introduction, we have come halfway in the analysis of the new iron industry.

In the discussion of what was being produced, it was convenient to treat the two branches of the iron industry in parallel, contrasting in succession their volumes of production, the demand curves they faced, and their supply schedules. The problems of organization are treated more easily separately, and the two branches of the new iron industry will be discussed consecutively. The description and analysis of the methods of production, of course, appear in both discussions, forming the bridge between the other two concerns. This chapter, then, discusses the changes in the size, integration, and technology of blast furnaces between 1830 and 1865 with the aim not only of describing these changes, but also of ascertaining at least a few of their causes and effects. Chapter 5 offers a similar analysis of rolling mills.

In the years preceding 1865, blast furnaces were less dependent upon the refiners of iron for their demand than at any time before the nineteenth century. The last third of the nineteenth century would see their freedom encroached upon by the producers of wrought iron and of steel, but for a few decades the high demand for cast iron, together with the failure of the rolling mills to supply rails and the American capital market to supply capital to the railroads, kept the blast furnaces of this country relatively, but by no means completely, independent of the demand for more advanced forms of iron.

In this time, the blast furnace began the transition from medieval technology and organization to more modern forms of production. Charcoal was replaced by mineral fuel, and the semifeudal iron plantation was replaced by the urban establishment and the company town. The proportion of pig iron made with charcoal declined from close to 100 per cent in 1840 to about 45 per cent in the middle 1850's, when the AISA data begin, and to about 25 per cent at the close of the Civil War.[1] The start of this trend took place under the umbrella of a rapidly rising level of total output of pig iron, while the decade before 1865 saw little rise in this level. The movement toward mineral fuel, therefore, was initiated by the addition of more mineral fuel furnaces than charcoal furnaces in the 1840's and carried on by the elimination of more of the latter than the former in the 1850's.

The rise of mineral fuel began in the East, and was confined almost solely to that region in the 1840's; the fuel in use was anthracite. In the succeeding years, the railroads and the Civil War increased communication between the East and West and facilitated the elimination of charcoal as a fuel, increasing the share of the market supplied by anthracite pig iron. In the expansion of production following the Civil War, coke would be the leading fuel, as the West began to take the lead in technology from the East. But in the period now under discussion, the East was the leading area of the industry and anthracite was the dominant fuel.

Although we will talk here of the change from the conditions of 1830 and before to the state of affairs in 1865 without detailing the manner of transition, this geographical bias must be remembered. For the movements to be discussed took place more completely and sooner in the East than in the West, and there were many areas of this country that did not feel the full strength of the forces under discussion until after the Civil War. This does not deny that leadership was occasionally to be found in

[1] Appendix C, Table C.3.

the West; it says that the *average* establishment in the East was more advanced than its counterpart in the West.

Before 1830, almost all the blast furnaces in the United States used charcoal for fuel. Charcoal was produced by exploiting extensive tracts of woodland and could not be transported with facility; as late as 1874, it ordinarily was transported only two to five miles.[2] The extensive nature of the production of charcoal and the difficulty of transporting it dictated a rural, isolated location for the blast furnaces using charcoal. The nature of the location dictated in its turn the use of a plantation system. As a result, iron plantations were the only sources of pig iron in the eighteenth century, and they lasted as long as charcoal was used for fuel.

The size of iron plantations varied widely within the constraints of technological necessity, but it was not uncommon in 1830 to find iron plantations of 10,000 acres or more.[3] We may calculate from technological factors why the area needed was so extensive. A series of technological transformations are available to us that give the area of woodland necessary to supply a blast furnace for a span of years, and allowances for other types of land bring the size of the plantation up to a range in the thousands of acres. One acre of timber provided about 30 cords of wood (on average), and each cord of wood yielded 40 bushels of charcoal (again on average). An acre of timber then supplied 1,200 bushels of charcoal. It took 180 bushels of charcoal to make a ton of pig iron from iron ore, which means that the wood from an acre of land would supply fuel to make $6\frac{2}{3}$ tons of pig iron. If the average production of a charcoal furnace is taken as 1,000 tons a year (which is accurate for the 1850's, but an overestimate for the eighteenth century), we find that 150 acres were necessary to supply the annual needs of a furnace. A stand of timber could regenerate itself in 20 years, and the requirement of a continuing supply of charcoal implied ownership of approximately 3,000 acres of woodland. Land was also needed for agricultural purposes, for the furnace and buildings, for working space, and to communicate between the not necessarily contiguous areas of usable

[2] I. Lowthian Bell, *Notes of a Visit to Coal- and Iron-Mines and Ironworks in the United States* (second edition; Newcastle-on-Tyne, 1875), p. 7.

[3] U.S., Congress, "Documents Relative to the Manufactures in the United States" (1833), II, 239-393.

timber. An average plantation size of somewhat larger than 3,000 acres is implied.[4]

In addition to the requirement for charcoal, the blast furnace needed water. Water is often neglected as a locational influence despite the fact that (like charcoal) it was not easily transportable, since (unlike charcoal) it was a free good where it was available. Yet water supplied power in the early years and the means of communicating power when steam was introduced. It supplied the essential cooling agent around the furnace, where production was a result of combustion. And it provided transportation for the finished goods in many cases. Maps of the early iron industry show furnaces scattered at the headwaters of many of the streams in Pennsylvania, especially those which fed into rivers serving the large cities such as Pittsburgh or Philadelphia.[5]

The organization of the iron plantations of the North was very similar to that of their somewhat later cousins of the South. They conducted a variety of activities designed to maintain the population of the plantation in addition to producing the commodity to be marketed. The ironmaster occupied a high position similar to that of the plantation owner of the South and often owned some of his laborers. The word "feudal" has been applied to the organization of the iron plantations, and the symbiotic relationship that is implied is perhaps accurate. The workers contributed their labor, while the master added the tools of enterprise and, instead of the medieval requirement of physical protection, the more modern benefits of communication with the production of other areas. In addition, the tone of patriarchal care that has survived from these days is reminiscent of an older form of organization.[6]

[4] The ratios used in this calculation were taken from Bell, pp. 6-7; Frederick Overman, *The Manufacture of Iron* (third edition; Philadelphia, 1854), p. 84, and Chapter 3. Louis C. Hunter, "The Heavy Industries Before 1860," in Harold F. Williamson (editor), *The Growth of the American Economy* (New York: Prentice-Hall, 1944), estimates the average size to be between 2,000 and 5,000 acres in 1850 (p. 176). Arthur Cecil Bining, *The Pennsylvania Iron Manufacture in the Eighteenth Century* (Harrisburg: Pennsylvania Historical Commission, 1938), lists several plantations for the eighteenth century with areas from 1,000 to 10,000 acres (p. 31).

[5] For example, the maps contained in J. P. Lesley, *The Iron Manufacturer's Guide to the Furnaces, Forges and Rolling Mills of the United States* (New York, 1859).

[6] Bining, *The Pennsylvania Iron Manufacture . . .*, pp. 29-48.

This tone may be suggested by a passage from Swank, itself reminiscent of a different way of looking at the world:

> Their [the iron plantations'] owners were almost feudal lords, to whom their workmen and their workmen's families looked for counsel and guidance in all the affairs of life as well as for employment; whose word was law; who often literally owned their black laborers, and to whom white "redemptioners" were frequently bound for a term of years to pay the cost of their passage across the ocean; who cultivated farms as well as made iron; who controlled the politics and largely maintained the churches and schools of their several neighborhoods; who were captains and colonels of military organizations; whose wives and daughters were grand ladies in the eyes of the simple people around them; whose dwellings were usually substantial structures, which were well furnished for that day and ordered in a style of liberality and hospitality.[7]

The end of the iron plantation was the result of the shift in blast-furnace fuel from charcoal to mineral fuel. Hunter remarks that if it took 2,000 to 5,000 acres to supply charcoal for a blast furnace, it nevertheless took only a six-foot seam covering half an acre to supply a furnace with coal.[8] The main reason for the rural location of blast furnaces disappeared, and the blast furnace of the "new" iron industry moved closer to other establishments and to coal deposits. The highly concentrated anthracite deposits of eastern Pennsylvania were the centers of the new iron industry.

This shift in location may have meant a lessening of the ironmaster's control over his workers and a decreased variety of operations, but it did not mean a smaller working force for an establishment. It was noted in Chapter 3 that the minimum economic size of a mineral fuel blast furnace was considerably larger than a charcoal furnace. The increase in output per worker at the new furnaces was offset by an increase in the output, and employment per furnace actually rose. In addition, the new locational influences encouraged the grouping of furnaces as much as the older technology had discouraged it, and this factor contributed to a growth of the size of iron establishments.

---

[7] James M. Swank, *History of the Manufacture of Iron in All Ages* (second edition; Philadelphia, 1892), p. 189.

[8] Hunter, "The Heavy Industries before 1860," p. 176.

The work force attached to a charcoal blast furnace did not change very much during this period. The Census of 1820 shows that most pig iron firms of that date had less than 100 employees, but that the firms did not cluster around any particular size smaller than that.[9] The McLane Report yields similar conclusions, but the data are not complete enough for most regions to provide a size distribution of the blast-furnace population. The data on western Pennsylvania are good, and will be used here. The data on eastern Pennsylvania are very poor, the results, we are told, of a prior questionnaire sent out by the Friends of Domestic Industry requesting the same information as the Treasury circular. Ironmasters, it was asserted, refused to fill out the second questionnaire. The Friends of Domestic Industry are the source of our information on total production for the years covered by the McLane Report, and this allegation may be taken as supporting their estimates. Peter Townsend, who was a prominent member of the New York group and whose estimate of iron production was used by them, was responsible for the New York State iron data of the McLane Report. The data are neatly tabulated, but they show a blast-furnace production of under 5,000 tons in 1830 and a foundry production of over 14,000. These two figures are incompatible, even without taking into account the other uses of iron in 1830, and Townsend's data must be used with care.[10]

For western Pennsylvania, the McLane Report notes the employment of 35 blast furnaces, averaging 54 employees (and 36 horses and oxen) apiece.[11] This may be compared with the data of the 1849 convention. The size of the establishment had grown during the intervening years, but not much; the average employment in 1849 was 65 men (and 41 oxen, horses, and mules) apiece, for charcoal furnaces in the West. There was no difference between the average employment of Eastern and Western charcoal furnaces, but hot-blast furnaces tended

[9] R. N. Grosse, "Determinants of the Size of Iron and Steel Firms in the United States, 1820-1880," unpublished doctoral dissertation, Harvard University, 1948, p. 55. Grosse's data come from only four states: Pennsylvania, New Jersey, Ohio, and Maryland. These states contained a large proportion of the iron industry, however, and general conclusions may be drawn from them.

[10] U.S., Congress, "Documents Relative to the Manufactures in the United States," II, 115-122, 198. See Appendix A for a discussion of Townsend's correction of the 1830 data on the production of iron.

[11] U.S., Congress, "Documents Relative to the Manufactures in the United States," II, 638-645.

to use more employees than cold-blast. In the East there were also many anthracite furnaces, with an average employment of about 80 persons per furnace and, because there were many multi-furnace firms, over 100 employees per firm.[12]

These data, unhappily, conflict with the Census of 1850. The Census shows an average employment per pig iron firm of 54 persons, which is the same as that shown in the McLane Report for 1830.[13] But it is not possible for an average of 65 and 100 or more to equal 54; the two sets of data for 1850 are not compatible. The conclusion that the employment of charcoal furnaces increased little or not at all in the years preceding 1850 is nevertheless supported by both sets of data, and we may neglect the conflict.

The Census of 1860 shows an average employment for blast-furnace establishments of 55 persons, unchanged from 1850.[14] This is again inconsistent with other data, as the continuing shift toward mineral fuel firms should have resulted in a rise in employment per firm. Grosse, however, has extracted from the original census returns for 1860 results which agree with the other data available. He found that the average employment of firms producing pig iron was about 100 persons. Charcoal-using firms had an average employment of slightly over 50, while mineral fuel firms were considerably over 100.[15] The Census of 1860 is therefore consistent with our other information, and the conclusion that the employment of charcoal firms stayed constant while mineral fuel firms had much larger, and possibly increasing, employment is supported. The published data of the Census do not bring this out, and it may be presumed that the raw 1850 Census returns are also not inconsistent with the other data from that time. The cause of the discrepancy, however, remains obscure.

The labor of the 50 or more men employed at a blast furnace may be described with perhaps a similar degree of exactitude as was shown in the discussion of their number. Most of these men were unskilled laborers doing jobs relating to the blast furnace or the auxiliary activities of the firm. Most wages at

---

[12] Convention of Iron Masters, *Documents Relating to the Manufacture of Iron, Published on Behalf of the Convention of Iron Masters which Met in Philadelphia on the 20th of December, 1849* (Philadelphia, 1850), Tables.

[13] Seventh Census, p. 181. The average for Pennsylvania was 52 persons.

[14] Eighth Census, p. clxxx. The Pennsylvania average was 61.

[15] Grosse, pp. 153-154.

an iron plantation in 1830 were about $20 a month, but skilled workers were paid piece rates and rather more.[16]

A furnace worked 24 hours a day, and the workers at the furnace were divided into 2 twelve-hour shifts of approximately 10 men each. The founders and the potters were the only skilled people at the furnace; they were in charge of castings made directly from the furnace, the potter making hollow ware and related fine castings. The other workers would "roast" the ore if it was necessary to eliminate sulphur, bring the materials to the furnace, charge the furnace, arrange for the tapping of the furnace and the making of pigs. The charge was measured by putting it in baskets, and this crude operation may have retarded the introduction of labor-saving devices for charging the furnace. Weighing the charge was introduced with anthracite and removed this difficulty.[17]

The other employees of the furnace were connected with auxiliary operations. The raw materials had to be mined or cut, and transported to the furnace. The finished product had to be taken to the point of sale and new provisions brought in. The large number of animals used at the blast furnaces of this period provided transport; they required several men for their care and handling. And at a plantation there were many other tasks necessary for the maintenance of the community.

Ore and coal were mined in a similar fashion. The work was usually done in an open pit or by means of simple shaft or tunnel operations. The materials were still sufficiently accessible so that although it was necessary to be cognizant of the difficulties of mining in special circumstances, it was seldom necessary to use this knowledge. Mining was done with pick and shovel, and power was used only for hauling and occasionally for ventilating operations. The scale of mining operations was growing, but in this period the growth had not made many inroads on the ironmaster's incentives to own his raw materials.[18]

An eighteenth-century blast furnace could be supplied by three or four ore miners, and a dozen or more woodcutters and "coalers" to prepare the charcoal.[19] These requirements

[16] U.S., Congress, "Documents Relative to the Manufactures in the United States." II, 255-391.

[17] Bining, The Pennsylvania Iron Manufacture . . ., pp. 79-81, 119-121.

[18] R. C. Taylor, Statistics of Coal (second edition; Philadelphia, 1855), p. 95; Overman, pp. 52-68, 92-99.

[19] Bining, Pennsylvania Iron Manufacture . . ., pp. 70, 73.

were undoubtedly increased by the growth of furnaces in the nineteenth century, but the order of magnitude suggests the small scale on which these operations were carried out. The woodcutters and charcoal coalers were, of course, being replaced by coal miners and, in the West, cokers. The coking of coal was done initially in the same way as charcoal was made from wood, the process being to burn the wood or the coal in the absence of air to expel the gases and leave the carbon. This was done in heaps covered with sod or dust, although coke soon began to be made in ovens. These ovens were of various design, their purpose being to protect the coking operations from storms. It was not till later that they were used to retain heat or to provide by-products.[20]

The preceding pages have shown that the size of the workforce at a blast furnace using a particular fuel and the nature of its work did not change much in the years before 1865. The investment per furnace also stayed roughly constant for each type of fuel, and the investment for mineral fuel furnaces, like the workforce, was larger than that for charcoal ones. This meant that the average investment represented by a blast furnace rose, and the tendency of firms operating mineral fuel blast furnaces to have more than one furnace meant that the investment per firm rose even faster.

Grosse shows that the capital of iron firms reported by the 1820 Census ranged from under $1,000 to $220,000. The mean was about $30,000, but incomplete coverage and the inclusion of forges and bloomaries in these data make this figure only an approximation.[21] The McLane Report gives an average investment of just over $30,000 for the 35 furnaces listed in western Pennsylvania, with a range from $5,000 to $110,000.[22] These figures are remarkably similar to the $33,000 average investment for charcoal blast furnaces in western Pennsylvania supplied by the 1849 convention of ironmasters. In the convention's data, however, we find that the charcoal blast furnaces in eastern Pennsylvania had a somewhat larger average investment, slightly over $40,000, which may cast doubt on the generality of the 1830 data. The convention also noted that

[20] Overman, pp. 119-129.
[21] Grosse, pp. 91-92.   .
[22] U.S., Congress, "Documents Relative to the Manufactures in the United States," II, 638-645.

furnaces using anthracite had an average investment of $56,000, and that the average investment of the firms owning these furnaces was $83,000.[23] The Census of 1850 reflected the shift to mineral fuel only slightly, and it showed an average investment per blast-furnace establishment to be $46,000. But by 1860, the shift to mineral fuel had advanced to such a degree that the average investment per firm was $81,000, and Pennsylvania had a per firm investment of $102,000.[24] These figures may suffer from the same faults that produced inconsistencies in the employment data, and they must be treated with caution. If they are correct, and the effects of the slightly higher price level in 1860 than in 1850 may be neglected, they imply that the size of investment in mineral fuel furnace firms was rising, even if the size of charcoal-using firms was not.

Integration

The firms that operated blast furnaces were, then, among the largest of their day, whether measured by number of employees, amount of fixed capital, or, in the days of charcoal, size of land holdings. These firms have been termed "independent" for this discussion, but they were independent only in a very restricted sense. They were independent of the demand for wrought iron in the sense that not much of their product, compared to other eras, was used by the refining sector of the industry. But the failure of pig iron production to expand greatly during the decade preceding the Civil War must be attributed at least partly to the vagaries of the demand for rails, the most important single component of the demand for wrought iron. Independence may be taken to refer mainly to the organizational structure of the industry, rather than the economic. Thus, without attempting to deny the existence of economic interdependence between blast furnaces and other economic entities, we may assert that the control of these various entities, in an organizational sense, was vested in different, independent people. In simple terms, the iron industry was not an integrated industry at this time.

[23] Convention of Iron Masters, *Documents* . . ., Tables. These figures were presented in Chapter 3 as investment per ton of capacity, rather than per firm, as here.

[24] Seventh Census, p. 181; Eighth Census, p. clxxx.

The separation between blast furnaces and forges or rolling mills does not imply that blast furnaces were not integrated backward. The industry was divided into two halves along technological lines, but within the smelting half there was integration, as is obvious from the existence of iron plantations. As the plantations disappeared, the structure of the industry changed, and blast furnaces became less integrated backward and more integrated forward. The former was the result of the shift in inputs for the blast furnace; the latter the result of the shift in demand toward wrought iron and the introduction of large-scale methods of producing it. We may consider the causes of these changes and also ask whether they had advanced far enough before the end of the Civil War for their effects to be noticeable in the industry's point of view.

The shift away from charcoal as a blast-furnace fuel created the opportunity for centralization in the supply of fuel. This was paralleled somewhat later by the introduction of distant, and centralized, sources of ore in Missouri and Michigan. These two changes tended to reduce the backward integration of the blast furnace, that is, the extent to which the blast furnace owned its own raw materials supplies. The mechanism by which they acted depended upon the twin factors of location and scale of production of the raw materials.[25] The ironmaster who made his own charcoal and mined his own ore was adding to his main activities small-scale, hand operations which had to be carried out on an extensive basis or in several units to supply his needs. The costs of transport dictated a grouping of these functions around the blast furnace, and it was natural for the ironmaster to control the ancillary operations. He would not try to control a larger operation, or one that was carried on at a great distance at this time, but small close operations came under his jurisdiction. The primary motive was control, the assurance of an adequate supply at a satisfactory quality of these materials. It was not enough to rely on a few people, and the markets for charcoal and iron ore were not well enough developed to assure him an adequate supply at all times. The necessity of maintaining the blast furnace in blast meant that it

[25] These ideas were suggested by the discussion of Grosse: For example, pp. 170-172, 277. They may have been suggested to him by the comments of Louis C. Hunter, "The Influence of the Market upon Technique in the Iron Industry in Western Pennsylvania Up to 1860," *Journal of Economic and Business History*, 1 (February, 1929), 280.

was costly to run short of supplies. The ironmaster consequently owned and controlled his raw materials.

The growth of mineral fuel production changed this state of affairs. The location of coal mining was centralized and the scale, while not large initially, grew over time as capital requirements increased and mining acquired economies of scale. Location was the most important influence on integration before 1865. The concentration of coal production to a degree that could not be matched by consumption implied the growth of a market in coal. If the market grew and competitive conditions were established, there was no incentive for the ironmaster to produce his own coal. Only when the economies of scale of mining became sufficiently large to prevent a competitive price from emerging was there an incentive for the ironmaster to go back into the production of coal.

But the disassociation of iron smelting from the production of the raw materials proceeded slowly. The constant size of the blast-furnace firms indicates that they did not contract the scope of their operations when they changed to coal. And the long discussions about the mining and treatment of coal in the iron manuals of the day indicate that the typical blast furnace owned its supply of raw material. There were, of course, firms which led the trend. Anthracite was developing into a widely sold fuel, and anthracite mining was developing into a competitive industry. Consequently, some of the ironmasters of that region bought their fuel rather than produce it. A precise dating of this movement into the market for coal cannot be made, but by the time of the depression of the 1870's the ironmasters were buying most of their coal from outside suppliers. Even if the movement was not well advanced in the ante bellum period, it must have already begun.

The movement of control in the production of iron ore can be traced more exactly, for here the locational factor was the only one operative. There was no difference in mining technique at this time according to the location or type of ore. A difference in technique would be one of the prime inducements for exploiting the Lake Superior ores, but mechanized iron mining was introduced after the Civil War. It was location, the distance between ores and the furnaces, together with the centralization of production in large-scale mines, that brought iron ore onto the market. The first ore deposits of national significance were

the ones at Pilot Knob and Iron Mountain in Missouri, discovered in the 1830's, and introduced into the general commerce of the Ohio River in the following decade. By the middle
1850's they were well known, and had brought the concept
of transporting ore over long distances into the realm of serious
consideration.[26]

By the time the Missouri ores had become famous, a new
source of ore had been discovered and its exploitation was
beginning. Ores in the upper peninsula of Michigan were
discovered by people interested in mining in the course of
surveying operations in the middle 1840's. They were not commercially exploited on a regular basis until a decade later,
due partly to the difficulties of transportation. Railroads were
needed to get the ore to the lake front, and a canal was needed
to eliminate transshipment at the falls opposite Sault Ste. Marie,
between Lake Superior and the other lakes. Railroads were
built piecemeal as various mining companies were organized
and entered the market, and the canal was initiated by a fed-
eral land grant to the State of Michigan. The state contracted
the work to private parties and the canal was first opened for
traffic in the summer of 1855. The first shipments of ore from
Lake Superior were made in 1850; the first commercial shipments, in 1854; and the first sizable shipments, coming through
the canal, in 1856.[27] These shipments were still small — 130,000
tons of iron ore were mined in Michigan in 1860, while 2,310,000
tons were used by blast furnaces —[28] but they would rise until
the "lake ores" dominated the industry in the latter years of
the nineteenth century.

Backward integration, then, was almost universal among blast
furnaces, even though it was declining. Forward integration
was the exception rather than the rule, but it was on the increase.
The iron plantations sometimes included a forge to convert the
pig iron produced into wrought iron, but the necessity of supply-
ing charcoal to the forge meant that it had to be located apart
from the blast furnace and there was little economy to be gained
by having both operations on one plantation. In the days when

[26] Swank, p. 335; Fritz Redlich, *History of American Business Leaders* (Ann
Arbor, Michigan: Edwards Brothers, 1940), Vol. I, p. 142; *Hunt's, 32* (1855), 514.

[27] Swank, pp. 320-331; A. P. Swineford, *History and Review of the Copper,
Iron, Silver, Slate and Other Material Interests of the South Shore of Lake
Superior* (Marquette, Michigan, 1879), pp. 90-131.

[28] Eighth Census, pp. clxxvi, clxxx.

iron was used primarily to make wrought iron, there was an incentive for ironmasters to own forges, which were smaller than blast furnaces, to use their product. Before pig iron became a well-traded commodity in its own right, however, much of the production of iron was made directly into the form of wrought iron and the problems of integration did not arise.

Grosse found that less than 20 per cent of the blast furnaces he studied in 1820 were integrated with forges. This proportion is echoed by the McLane Report's data on western Pennsylvania, although the number of firms listing both furnaces and forges without showing production of both pig and bar iron was somewhat higher. Firms studied for 1857 show approximately the same proportion of integrated (forward) firms.[29]

The most obvious cause of the separation between the two branches of the iron industry was the demand for pig iron in unconverted form to be used in castings; location and the shortage of capital for the charcoal iron industry may explain much of the rest for the early years. But the growth of rolling mills began to create a new situation about 1850. They were much larger than the forges which they replaced, and the forces that drive the larger members of a market situation to try to control the smaller member would be expected to operate. These forces will be considered in more detail in the following chapter. Here we may note that there was a split between the size of rail mills and other rolling mills, by which the forces operated most strongly upon the small, but growing number of rail mills, and that a trend toward integration was resulting from the change from forge to rolling mill. Charcoal furnaces supplied the forges with iron to refine; mineral fuel furnaces supplied the rolling mills. The degree of integration among mineral fuel blast furnaces was larger than that among charcoal blast furnaces.[30]

The change to mineral fuel, then, appears to have had only a restricted effect on the organization of the smelting branch of the iron industry by 1865. The preceding pages have shown that the extent to which the blast furnaces were part of integrated firms had not changed greatly, although they had begun to become less integrated backward and more integrated for-

[29] Grosse, p. 270; U.S., Congress, "Documents Relative to the Manufactures in the United States," II, 638-645.

[30] Grosse, p. 278. Only 13 per cent of the output of charcoal pig iron was produced by integrated firms, while 26 per cent of coal pig iron was so produced in 1857.

ward. The nature of competition in the markets for raw materials or for pig iron had consequently not changed greatly either. Most blast furnaces were not dependent upon the market for their ore or fuel, and they usually sold their products in a competitive market. The blast-furnace firm remained a large organization, and there is little reason to suspect that the shift away from a completely isolated, rural environment had much effect on the running of the firm, even though it may have restricted the range of the ironmaster's feudal activities.

Even though the structure of the industry and of the market for pig iron had not changed, we may note a significant change in the extent of the market. The growth of railroads, and particularly the construction of through lines between the East and the West, meant that there was more nearly a national market in iron. The large weight of a dollar's worth of iron meant that there were always local markets and advantages accruing to the closest producer, but the cheapening of transport reduced the extent of local monopolies and increased the competition between producers of various regions. In economic terms, the cost of transporting goods across natural barriers had declined sufficiently to make the location of these natural barriers only one determinant among many of the selling range of producers. The most obvious result of this increase in movement was the high mortality rate among Western charcoal iron producers in the 1850's, when, aside from the transitory peak in 1853-1854, iron prices failed to maintain the level of the previous decade.

Two things are worth noting about this movement. First, the price in a region depends upon the price of the marginal commodity. It is not necessary for large amounts of a good to be transported from East to West for it to affect the prices in the West. If the potentiality is there, it will keep the Western price down, even though almost all the goods sold in the West were made there. Second, the effect of the larger size of the new furnaces and the stagnant level of production in the 1850's was to decrease the number of establishments making pig iron. The Census of 1860 shows a smaller number of blast-furnace establishments than the 1850 Census, even though the production of pig iron was much greater in the former year than in the depressed conditions of the latter.[31] This problem was the result of the change in firm size and the small secular increase

[31] Seventh Census, p. 181; Eighth Census, p. clxxx.

of total production; it did not depend upon the foresight or lack of it exhibited by ironmasters. Some of them were to be forced out of the market, and the extent to which knowledge was disseminated might determine who survived, but it could not affect the number of survivors.

Technological Change

The attitude of the new ironmasters toward innovations in this period was not different from that of their predecessors due to the influence of industrial organization, as the organizational changes in the pig iron industry that finally resulted from the change in fuels and in demand had not yet taken place. (This is not true of the wrought-iron branch of the industry, and the role of industrial organization will be more fully explored there.) The new innovations that were adopted by blast-furnace operators in the years before 1865 were mostly adjuncts to the use of mineral fuel and they were used primarily by the mineral fuel firms. Innovations of a more general nature, leading to increased output or decreased costs of materials handling, were not widely used before 1865. Their use after that date was sponsored by the integrated blast furnaces of that era, and the factor of industrial organization will be discussed in that connection.

In the period preceding 1865, only a sequence of operations designed to facilitate the use of mineral fuel, or flowing in an obvious way from its use, was the subject of interest. It will be recalled from the discussion of Chapter 3 that the success of the new fuels depended on auxiliary innovations that increased the scale of the blast furnace, primarily a hot and powerful blast. The heat content of the waste blast–furnace gases was used to heat the blast, and ironmasters soon discovered that it could also be used to provide the pressure. It was a short step from that to the use of waste gases to supply power for operations in general and for the handling of materials in particular, but this step had to wait until after 1865 in all but a few exceptional cases.[32]

The combination of hotter and more powerful blasts was the force making for the increased output and economy of blast

[32] A few furnaces used mechanical lifts to get the charge to the top of the furnace, and these lifts presumably used steam power generated by waste gases.

furnaces in this period; a cheapened supply of these two qualities meant a more efficient blast furnace. But the blast was
not heated as high as it would go later, nor was it subject to
as much pressure as it would later receive. This was partly
the effect of an insufficient technology, but there was little
desire for the creation of this technology until after the Civil
War. It was enough to absorb the new fuels and their immediate
implications at this time. Consequently the auxiliary operations
were not altered to any great extent until after 1865.

The progress of the hot blast has been chronicled already.
Its further progress through increases in its temperature will
be explored later for the period in which it is important. In
the 1850's the temperature of the blast at the most advanced
furnaces was regulated by keeping it between the melting point
of lead and of zinc, that is, between 612° and 787° F. Any higher
temperature was thought to "burn" the iron.[33] The thought of
increasing the output of the furnace by changing the character
of the blast alone had not penetrated the industry, despite its
similarity to the problem of the introduction of mineral fuel.

Waste gases were used to heat the blast almost universally.
Pipes containing the blast were run through "ovens" at the
top of the furnace where gases from the furnace communicated
their heat to the air in the pipes. The pipes were made of cast
iron, and one of the limitations on the temperature of the blast
was the durability of the pipes at high temperatures.[34] The use
of closed tops on the blast furnace and "downcomers" to take
the blast to the ground where it was used for heat were new,
and not enthusiastically received. Suspicion was high that the
closed top affected the iron, as the hot blast had done earlier.
In the 1850's, this was an issue that was being discussed and
twenty years later there was still enough interest in the subject
to inspire an article denying the bad effects that had been previously attributed to the closed top.[35] Needless to say, closed
tops were the exception before 1865. Even the use of waste

[33] Joseph G. Butler, Jr., *Fifty Years of Iron and Steel* (fourth edition; Cleveland:
The Penton Press, 1923), p. 14; *Bulletin, 3* (February 17, 1869), 185.
[34] Overman, pp. 428-442; John Wilson, *Special Report on the New York
Industrial Exhibition*, British Parliamentary Reports, 1854, Vol. 36, p. 36.
[35] JFI, 3rd series, 25 (1853), 188-193, 243-247; Frank Firmstone, "Comparison
of Results from Open-Topped and Closed-Topped Furnaces," *AIME, 4* (October,
1875), 128-132; John B. Pearse, *A Concise History of Iron Manufacture of the
American Colonies* (Philadelphia, 1876), pp. 160, 247.

gases was not universal. There were many charcoal furnaces that did not use this form of heat or of power; they did not use a hot blast, and water provided their power.[36]

The main use, if not the only use, of steam *power* at the blast furnace was to run the blowing or blast engines. These engines had increased greatly in size from the wooden tubs used to power the blast in the years around 1830, although their size was not at all standardized. In the late 1850's, an advanced furnace might use a 500-horsepower engine to supply the blast for two furnaces, or 160 horsepower to run four blast furnaces. It was not uncommon for a firm to find that it was better to use the power it had formerly used for three furnaces to run two and to build a new blast engine for the third.[37]

The state of blast engines of the ante bellum era can be seen from the words of one veteran of the industry: "If you reflect that the blast in those days was blown usually by an engine that had been worn out on a Mississippi River steamboat, and that it was the usual thing for the men about a furnace to operate the walking beam when the engine broke down, you will have some light on the strength and steadiness of the hot blast of that day."[38]

Nevertheless, anthracite had been introduced, and a hot, powerful blast was in use. And although the use of anthracite would be neglected in favor of processes involving coke, after 1865, blast-furnace blasts would become hotter and more powerful. These changes were related — developments in one area of technology forcing adaptions in others.[39]

[36] This statement cannot be documented, because the attention given to backward establishments is always minimal and no one would note explicitly that they were not using waste gases. The restriction of the use of waste gases to the few specific purposes mentioned, however, together with the continuing discussion about their usefulness, indicates that their use was far from universal.

[37] *Hunt's*, *34* (1856), 126; *39* (1858), 633; Trenton Iron Company, *Annual Report of the Secretary*, 1859, p. 7.

[38] Butler, p. 14.

[39] See Chapters 8 and 9.

5

The Iron Rolling Mills

Puddling and Rolling

In the years following 1830, the refining branch of the iron industry underwent a change similar to that undergone by the makers of pig iron. The change in the refining of iron, however, progressed much faster than the change in smelting iron. In the latter, as we have seen, just over half the industry had switched to the new methods by the middle 1850's; in the former, as we shall see shortly, approximately 90 per cent had changed by the same date. The change in blast-furnace technique was retarded by the reluctance of Western ironmasters to adopt new techniques, the conversion in the East going ahead with great rapidity. In the refining half of the industry, the West kept pace with the East and even may have led it slightly, a development which invalidates arguments explaining the retardation of the use of coke in the blast furnace by the "backwardness" of the West.

This chapter provides some of the details about the transformation of production in the refining branch of the industry, discussing both the introduction of the new methods of puddling and rolling and their improvement over time. The effects of changes in the methods of production on the size and integration of the industry are then discussed in a similar fashion to the treatment of blast furnaces in Chapter 4. The refining branch of the new iron industry differed from the smelting branch, however, in that it produced a substantial quantity of a product—rails—whose production affected the nature of technological change in the industry. This chapter and Part I close with a discussion of the rail mills and the conditions created by them for the introduction of Bessemer steel.

99

The new method of making wrought iron from pig iron consisted of a package of innovations, similar to the package of
innovations that accompanied the introduction of mineral coal
as a blast-furnace fuel, and we may treat this package as a
single complex innovation. Puddling (or its modification known
as boiling), mineral fuel, and the use of rolling in place of
hammering to shape wrought iron all were introduced together.[1]
Rolling was introduced before puddling in some cases, but the
links between the two processes were so strong that a separate
treatment of the way in which the iron industry switched to
their use would be pointless, even if possible.[2] As it is easier
to trace the introduction of rolling mills than of puddling, this
is used as the key to this collection of innovations.

Data from 1830 almost ignore the small number of rolling
mills in existence at that time. The Friends of Domestic Industry did not mention them; the McLane Report mentioned only
a few, and they were almost all in Pittsburgh. Much of the
early discussion of rolling mills derived from the experience
of western Pennsylvania, and this fact alone prevents us from
labeling the West as a backward region.[3]

In 1849, about 80 per cent of the wrought iron made in Pennsylvania was made in rolling mills. Almost all of the forge
production was located in the East, the number of forges in the
western part of the state having fallen to three. Almost all of
the rolling mills reported that they had puddling furnaces,
and the rolling mills with puddling furnaces produced over 80
per cent of rolling mill output in the East and 90 per cent in
the West. In other words, puddling did not lag far behind
rolling in its spread by 1850, any early discrepancies having
been almost entirely eliminated. Of the rolling mills in exis-

[1] For a description of puddling, see Chapter 1. The variant of puddling known
as "boiling" is described in Frederick Overman, *The Manufacture of Iron*
(third edition; Philadelphia, 1854), pp. 265-271, and in John Fritz, *The Autobiography of John Fritz* (New York: John Wiley and Sons, Inc., 1912), pp. 50-51.

[2] This is not intended to apply to their initial introduction in the United
States, where the two may be separated. See Louis C. Hunter, "The Influence
of the Market upon Technique in the Iron Industry in Western Pennsylvania
Up to 1860," *Journal of Economic and Business History*, 1 (February, 1929),
241-281.

[3] The Friends of Domestic Industry's report talks only of forges. Data on
the early rolling mills come from U.S., Congress, "Documents Relative to the
Manufactures in the United States," II, 202, 638-639; Hunter, "The Influence
of the Market . . .,"; James M. Swank, *History of the Manufacture of Iron in
All Ages* (second edition; Philadelphia, 1892), pp. 227-231.

tence in 1849, 40 per cent had been built in the previous five years, during the boom of the forties, and about 20 per cent had been built before 1830, in both the East and the West.[4]

By 1856, the proportion of wrought iron made in rolling mills had risen to 95 per cent, of which almost all was made by puddling. The total output of wrought iron in that year was about 520,000 tons. Of this amount, 29,000 tons were made directly from the ore by bloomaries, 53,000 was made from pig iron in forges, and the balance was made from pig iron and scrap by puddling or boiling. Only part of the products of bloomaries and forges, however, was shaped by hammers and sold as bars. Bloomaries sent 22,000 tons of their product to rolling mills in the form of blooms (unfinished shapes), and forges sent 39,000 tons. The result of these transactions was that 95 per cent of the wrought iron sold by the industry to others had been shaped by grooved rolls, although only 85 per cent had been made by puddling or boiling.[5] The similarity of 85 and 95 per cent lets us talk of puddling and rolling in one breath.

We may motivate this discussion of technological change with a demonstration that it was indeed technological advance. The information used is a cross-section comparison for 1849, similar to that used in the discussion of blast furnaces. But while it was necessary there to translate the data into profitability terms in order to discuss the motivations of ironmasters, we may stop here with the efficiency measures. The argument can presumably be extended to profitability, and there is no reason to expect difficulties to arise. The rolling mills of 1849 had an output per employee that was twice that of forges, and an output per unit of capital half again as large as the figure for forges. The coefficients for rolling mills are the same in the East and the West, but the figures for forges come from the East where almost all the remaining forges in Pennsylvania were located.[6]

The question which does arise is why, if the lack of a suitable coke retarded its use in the blast furnace, did not this lack of

[4] Convention of Iron Masters, *Documents Relating to the Manufacture of Iron, Published on Behalf of the Convention of Iron Masters which met in Philadelphia on the 20th of December, 1849* (Philadelphia, 1850), Tables.

[5] J. P. Lesley, *The Iron Manufacturer's Guide to the Furnaces, Forges and Rolling Mills of the United States* (New York, 1859), pp. 760-763.

[6] Convention of Iron Masters, *Documents . . .*, Tables. See also Table 5.1.

coke retard the spread of puddling and rolling in the West?
The answer, as Hunter pointed out, is that the requirements
for a puddling fuel are less stringent than the requirements
for a blast-furnace fuel. In a blast furnace, all the materials
are put together and they communicate their impurities to
the final product. In a reverberatory furnace, the fuel is separated from the iron and there is less opportunity for the impurities of the coke to be transmitted to the iron.[7] There is
also not the stringent requirement that the coke have a tough
enough physical structure to support the weight of the blast-
furnace charge above it. Consequently it was not necessary
to wait for the discovery and exploitation of high-quality coal
beds for coke to be used in refining iron. Hunter had the great
insight necessary to see the different requirements in the two
branches of the industry, although the present study has emphasized the elimination of this technical barrier to the production
of coke pig iron more than he did.

Improvements in Rolling Mill Practice

The preceding pages have demonstrated that a substantial
part of the increased productivity of the wrought-iron producing
branch of the new iron industry came from the introduction of
puddling and rolling. A further increase in productivity was
gained through the improvement of these processes themselves.

The principal innovations in puddling may be noted briefly.
Iron bottoms had had to be introduced before puddling was
adopted in America, and other modifications continued to increase the efficiency of this process. Boiling increased the
range of materials that could be used, and double puddling
furnaces increased the output from a given amount of capital
by allowing two men to work one furnace. The rotary squeezer,
an American invention of the 1840's, allowed the output of
the puddling furnace to be easily squeezed of extraneous matter
and readied for rolling.[8]

Rolling, like squeezing, was a mechanical process, but it
involved a much larger array of machines. The data that have
survived in coherent form relate to only one or two of the
machines in use and it is difficult if not impossible to assess

[7] Hunter, "The Influence of the Market . . .".

[8] Overman, pp. 259-305, 341-344; John Wilson, *Special Report on the New
York Industrial Exhibition* (British Parliamentary Reports, 1854), Vol. 36, p. 45.

the economic importance of these machines. This discussion will soon concentrate on one innovation, the introduction of the "three-high mill," in the hope that it was not only the most important single innovation of the period, but that it was dealing with problems that were typical of the industry.

The improvements in rolling mill practice in the years preceding the introduction of Bessemer steel may be seen as either improving the quality of the product produced, or as improving the ease with which the existing products could be fabricated. There were many experiments that were tried to extend the range of iron products, several of which remained the province of the Trenton Iron Company for this period. The manufacture of wrought-iron beams and of special quality rails were two of them.[9] The beams were used in the construction of Cooper Union and other buildings in New York; the rails were used by a few railroads that were willing to pay the extra premium for quality.

Most railroads, however, had difficulty obtaining the quality of rails they desired at the desired price. The Philadelphia and Reading Railroad was able to purchase satisfactory rails from the Fairmount Rolling Mill in Philadelphia in 1857. The president of the rolling mills was subsequently elected president of the railroad, and when he tried to induce other rolling mills to make rails to his specifications he could not get a satisfactory price. The solution to the problem was for him to build another rolling mill, which he did after the Civil War.[10] Products were available for those who were willing to pay for them or make them, that is, the technology existed, but there is no evidence that the average rolling mill could make either beams or high-quality rails. The introduction of new techniques cannot be confused with their dissemination.

Most of the energy of ironmasters was directed to improving the ease of making the existing line of products, which of course also meant increasing the ability to make more exacting products. The most obvious way of doing this was to improve the general character of the machines used to drive the rolls. They could be made larger and more resistant to strain; they could be made more reliable to avoid the frequent stoppages

[9] Trenton Iron Company, *Annual Report of the Secretary*, 1859, p. 6; Allan Nevins, *Abram S. Hewitt* (New York: Harper and Brothers, 1935), p. 104.
[10] W. E. C. Coxe, "Endurance of Iron Rails," *AIME*, 5 (June, 1876), 107-114.

that were a feature of ante bellum rail mills. These aims were somewhat at variance and the increasingly strong machines tended to become increasingly complex and subject to more, rather than less, breakage. But the capacity to produce heavy rolled products increased, although not enough to induce American ironmasters to adopt mechanical auxiliary equipment. The iron was moved around the mill by hand throughout this period.

The handling of iron rails in the mills created a problem of cooling, and to discuss it we need a little technical background. A roll "train" consisted of two large, grooved rolls between which the iron was passed successively, through smaller and smaller grooves, until it was changed from a bar of iron to the finished rail. The bar to be rolled, or "pile" as it was called, was made out of many strips of wrought iron which were welded together under a hammer or by rolling. The rolls were operated by a steam engine and operated in only one direction, and the rail accordingly had to be passed back over the rolls each time to ready it for the next "pass." This operation took time and the rail would cool off. Cool rails became red-short, that is, brittle, and they would split in the rolls, the weld between the different bars in the rail pile often coming loose. Time was lost fixing the rail and the rail cooled off more. These difficulties created extra costs, in the form of extra work and heating, and they could also damage the rolling machinery. There was more than the usual incentive of saving capital costs to encourage increasing the speed of rolling.

The Cambria Iron Works were very prone to these difficulties, and although their difficulties and failures were attributed by some to the bad quality of Johnstown iron, they were manifested in the above ways and could be solved by better machinery.[11] The owners did not think so, but one of their employees did. He was John Fritz, one of the fine engineering minds of the iron and steel industry, who was brought to Cambria by David Reeves, a member of the family that owned several rival concerns. The reason for this odd behavior appears to be that the Reeves family had fingers in several pies, including Cambria.[12]

[11] See the discussion of the size and integration of the Cambria Iron Works, pp. 109-111.

[12] *Bulletin*, 25 (December 2, 1891), 352; 28 (October 17, 1894), 234; Fritz, pp. 90, 92, 100.

The English solution to these difficulties had been to adopt reversing mills. These mills had engines whose direction could be reversed, permitting the rail to be run through the rolls in both directions, and eliminating the time consumed in passing the rail back over the rolls. Announcements of the reversing mill were present in America as early as 1852, but they could not have been widespread, as Overman does not mention them in the third edition of his production manual, published in 1854.[13]

Fritz did not adopt this solution, and a later writer commented that the American preference for the three-high mill, Fritz's innovation, over reversing mills showed the spirit and competitiveness of ironmasters because of its greater possibilities for expansion.[14] Whether or not Fritz thought of it for this reason, he would certainly agree with this evaluation. When he came to Johnstown he found the Cambria Company under the supervision of Morrell, of whom he wrote later: "He was a very clever gentleman, but knew nothing of the iron business, which, to say the least, was unfortunate."[15] Cambria was laboring under many difficulties, and Fritz thought he could remedy them. When he proposed to rebuild the mill with newly designed and heavier machinery, the owners objected and forced the managers to take responsibility for all losses stemming from this experiment. The experiment was, of course, the three-high mill. The idea of putting an extra roll on top of the roll train so that the iron could be rolled on its way back as well as on its way forward was not a new one; this type of mill had already been applied to smaller iron products. But to apply it on the scale that Fritz was about to attempt was to be more adventurous than the owners wished.

Fritz prevailed and he was able to build his new mill, in which he eliminated the traditional safety devices to strengthen the machinery. The mill worked for two days and then burned down. It is not clear how everyone reacted to this event, or how the blame was allocated between acts of God and the evil effects of new machinery. But the mill was speedily rebuilt, of brick rather than wood, and the three-high mill became a

[13] *JFI*, 3rd series, 24 (1852), 143-144; Overman, pp. 344-371.
[14] Harry Huse Campbell, *The Manufacture and Properties of Iron and Steel* (second edition; New York: The Engineering and Mining Journal, 1903), pp. 603-614.
[15] Fritz, p. 107.

financial success. Cambria became the technological leader in the rolling of iron rails, and Fritz became the leading engineer of the industry.[16]

There are no figures on the increase in productivity or profitability of the three-high mill, but it was stated everywhere to be beneficial. One of its effects was to increase the capacity of rail mills, although there are no estimates of how much the cost increased at the same time. An obituary of John Fritz's brother George, with whom he worked at Cambria, asserted that the Fritz brothers had almost created the American rail mill and that the three-high mill made Cambria a financial success when rail making was a failure everywhere else.[17] This last statement must be taken with a grain of salt in view of the steady rise of rail production in these years, but the feeling expressed was quite general. By 1865, one-third of all rail trains (37 out of 131) were three-high and in the 1880's, they all were.[18]

Size and Integration

We have supported the contention that the supply curve for wrought iron was falling faster than the supply curve for pig iron by appealing to (qualitative) evidence on technological change. We now turn to the effects this change had upon the industry itself and to its attitude toward such changes. As with blast furnaces, the innovations in refining were beneficial through the economies of scale they brought with them. The result was a large increase in the size of the average firm. The production of wrought iron separated into two branches as the demand for rails induced ironmasters to invest in the even larger capital equipment necessary for the production of these large articles. The large size of the required capital created a concentrated subindustry, and the fortunes of this group began to diverge from those of the rest of the industry. One of the concerns of this discussion is to identify this difference of fortunes and to investigate, if possible, its causes. To anticipate our conclusion, there are several highly correlated

[16] Fritz, pp. 107-121; *Bulletin, 28* (October 17, 1894), 234; Overman, pp. 354-358. (For accounts of earlier three-high mills.)

[17] *E&MJ, 16* (1873), 153.

[18] Samuel Harries Daddow and Benjamin Bannan, *Coal, Iron, and Oil, or the Practical American Miner* (Pottsville, Pennsylvania, 1866), p. 683; *Bulletin, 16* (February 22, 1882), 57.

variables that could be used to explain the divergence — size, economies of scale, concentration — and little prospect of separating their influence. There is a pattern of accumulating economies of scale and their concomitant effects, but the causes of this self-perpetuating trend remain obscure.

The forges of this period did not grow in size, although there was a shift among forges from those that made wrought iron directly from the ore (bloomaries) to those that made wrought iron from pig iron (refinery forges). In the four states studied by Grosse, there were half as many bloomaries as forges in 1820. This implies that the proportion of production accounted for by bloomaries was less than one-third since bloomaries were somewhat smaller than forges; the estimates of the Friends of Domestic Industry presented in Chapter 1 would suggest that it was much less. In any case, the proportion of forges in these states that worked directly from the ore had fallen to about 15 per cent by 1860.[19] Forges in 1820 employed less than 30 people apiece as a rule, and bloomaries employed less than 20. The forges reported by the McLane Report appear to be slightly larger, but data for 1850 and 1860 agree with the 1820 estimates. Grosse even finds that the average size of independent forges fell between 1820 and 1860 as the larger forges were replaced by rolling mills.[20]

Rolling mills, which here play the role that mineral fuel furnaces played among blast furnaces, appear to have increased their employment over time, but the only observations that are large enough to be meaningful are from 1850 and later. A comparison of sizes as of that date will have to do.

There were, in 1849, the following number of wrought iron works in eastern Pennsylvania: 6 bloomaries, 118 forges, 56 rolling mills. There were in western Pennsylvania at the same time 3 forges and 23 rolling mills. The completeness of the transition to the new form of production in the West is apparent from these figures. In addition, the rolling mills of the West were almost all located in Pittsburgh, while those of the East were scattered rather widely. The average sizes of these works are summarized in Table 5.1.

The small size of forges is immediately apparent, by what-

[19] R. N. Grosse, "Determinants of the Size of Iron and Steel Firms in the United States, 1820-1880," unpublished doctoral dissertation, Harvard University, 1948, p. 267.

[20] Ibid., pp. 59, 158, 162, 256; U.S., Congress, "Documents Relative to the Manufactures in the United States," II, 120-121, 199, 638-645.

ever measure is used. Rolling mills were much larger establishments, and the Western rolling mills, which means mostly the
Pittsburgh mills, were much larger than their Eastern counterparts. A little division, however, shows that they were not
more efficient. The largest product per employee for rolling
mills in both regions was about 25 tons, and the largest product per $1,000 of capital was near 30 tons. This may be con-
trasted with the analogous figures of 13 and 20 for (Eastern)
forges, but the lack of difference between the efficiency ratios
for the different size rolling mills is surprising.

TABLE 5.1
AVERAGE SIZE OF VARIOUS IRONWORKS, 1849

	Fixed Capital (thousands of $)	Largest Product (tons)	Employment (no. of persons)
Eastern bloomaries	4.8	91	16
forges	17	360	26
rolling mills	56	1640	65
Western forges	4.7	127	14
rolling mills	105	3130	131

Source: Convention of Iron Masters, *Documents* . . ., Tables.

The different sources of power in use in the two regions
explains most of the discrepancy. The rolling mills in the
East were scattered over the eastern slope of the Allegheny
Mountains, where there was an abundant supply of water power.
The Western rolling mills were located near the water transportation system of the West which, by virtue of being suitable
for transportation, did not possess the rapid descents that
produce power. As a result, all but one of the Western rolling
mills listed by the 1849 convention used steam power, while
less than half of the Eastern mills used steam. Those mills
in the East which did use steam were larger than their water-
powered counterparts, having an average largest product of
about 2,800 tons compared with 860 tons for water-powered
mills.[21] The average size of rolling mills powered by steam
was not very different between the two regions, and the dif-

[21] Convention of Iron Masters, *Documents* . . ., Tables. I have included the
two Eastern mills using both water and steam power with steam-powered mills;
the comparison is with mills which used only water power.

ference may have been due to random factors. The use of steam, therefore, increased the efficient size of rolling mills without increasing their productivity at this new size relative to the productivity of water-powered mills at their most efficient size. The use of steam was a marginal change at this time, adopted or not depending upon the relative prices in a particular locale.

These data are plant data, but Grosse has compiled firm data for 1860. In these data it is necessary to differentiate between types of rolling mills to analyze their size. The firms that made rails had an average of about 400 employees, while other kinds of rolling mills had far less. Firms making nails employed an average of 175 persons, and firms providing the market with bars, sheets, plates, tubes, and so forth, had an average of less than 100.[22] Looking at plants, then, rolling mills were much larger than forges, and steam-powered rolling mills were the largest among rolling mills. Firms producing rails were the largest of the rolling mill firms, though, as they were most often integrated in addition to using steam.

We noted in the last chapter that integration in the iron industry had not proceeded very far by 1865. Here we may note that the leaders in the trend to integration were the rail mills. According to a British investigator, there were only six integrated ironworks in the United States in 1854. His list does not appear to be complete, but all but one of the firms he lists were engaged in making rails, and the other well-known integrated firms were also rail producers.[23]

A few of the largest iron firms are worth our attention, as they were controlled by some of the most well-known names of the iron industry. The four largest iron firms of 1860 were all integrated rail mills. The Montour Iron Works was the largest with 3,000 employees, and it was followed by Wood, Morrell and Company (the Cambria Iron Works) with 1,948, the Phoenix Iron Company with 1,230, and the Trenton Iron Company with 786.[24] Abram S. Hewitt, whose comments on

[22] Grosse, p. 152.

[23] Wilson, p. 43, The integrated nonrail firm was Fuller and Lord of Boonton, New Jersey, who made nails and spikes. Wilson does not list Cambria or Brady's Bend as integrated firms. The reason may be that they were in financial difficulties at the time.

[24] Grosse, p. 155. The Phoenix Iron Company was located at Phoenixville, Pennsylvania, and is often referred to by its location instead of its name.

iron affairs we have quoted and will quote again, was an owner and manager of the Trenton Iron Company. Under his management the company was always among the first to try new techniques and introduce new products. They were often, in fact, too early to make a commercial success of these experiments and the honor of initial commercial production usually belongs elsewhere. The wrought-iron beams referred to in Chapter 2 are a conspicuous exception. The Phoenix Iron Company was controlled by the Reeves family, as were several other ironworks, including the Safe Harbor Rolling Mill, also listed as an integrated ironworks in 1854. Samuel J. Reeves was one of the sources for the pig iron production estimates in the middle 1840's, and he was the inventor of the famous "Phoenix wrought-iron column," a structural shape for iron, in 1862.[25] Daniel J. Morrell was the general manager of the Cambria Iron Works; he has already appeared in connection with the three-high mill, and he will be a protagonist in the play acted out during the introduction of Bessemer steel.

Cambria first came into being as a unified entity in 1852, when a Boston group bought and put under unified control four charcoal blast furnaces located in or about Johnstown, Pennsylvania. Besides the natural advantages of the region in raw materials, the arrival of the Pennsylvania Railroad in Johnstown in 1850 undoubtedly recommended the property to the Boston men. The new owners soon began to construct four coke furnaces, taking the attitude that would be implied by the calculations of Chapter 3. But before they could finish the furnaces and put them into operation, they were overtaken by payments due on outstanding bills and forced to suspend. This was in 1854, and the Philadelphia creditors appointed Morrell to investigate and to discover the best means for preserving their investment. Morrell recommended further investment in the Johnstown company and the creditors agreed. Unhappily, this attempt was no more successful than the first, and the company was forced to suspend again in 1855. Morrell was by then involved with the company, and he persuaded a new group of Philadelphia people to lease the ironworks for seven years. In this fashion Wood, Morrell and Company was

<hr>

[25] Daddow and Bannan, pp. 694-695; *Bulletin, 13* (December 17, 1879), 324; Chapter 1, and Appendix C.

formed, and Morrell came to leave Philadelphia for Johnstown to manage an iron business about which he knew very little.[26]

It is not surprising that the men whose names have come down to us as important should be the owners and managers of the largest firms. What we must ask now is why these firms had attained their large size. There were not yet any large technical economies from integration. The iron was cast into pigs as it came from the blast furnace and could be transported easily in that form. In fact, the Trenton Iron Company's blast furnaces were fifty miles away from its rolling mill.[27] There were some economies to be gained from the use of waste blast-furnace gases to generate power for the rolling mill, but the discussion of the previous chapter has emphasized the tentative nature of the use of waste gases before 1865, and this could not have been a very persuasive reason. It, too, could not apply to widely separated plants. The market for pig iron was competitive in nature, with most of the blast furnaces selling their product on the open market. Monopolistic profits from the ownership of raw materials were thus not to be gained from vertical integration.

The motives for integration were not persuasive and it is not surprising that less than one-fourth of the pig iron produced in 1860 was made by integrated firms. On the other hand, the integrated firms were among the more successful firms in the industry, and this cannot be due to chance. There are three possibilities: the character of the men just described may have controlled their desire for integration and also their ability to succeed; integration may have been important for the motives of control discussed in the previous chapter for blast furnaces and may have enabled rolling mills to operate better; or integration itself may have been more or less accidental while the effects of it in itself may have promoted activities that led to a greater degree of success. We consider these in turn.

It is not one of the purposes of this history to assess the role of the individual in the determination of economic and other

[26] This account is taken from Judge Joseph Masters, "Brief History of the early Iron and Steel Industry of the Wood, Morrell and Company and the Cambria Iron Company at Johnstown, Pa." (typescript, December 14, 1914). This small typescript was made available to me through the courtesy of Elting E. Morison.

[27] Trenton Iron Company, *Documents relating to the Trenton Iron Company*, p. 11.

events. Accordingly, the first possibility may be subsumed under one or the other of the following possibilities. Integration was in itself either beneficial or not to the ordinary running of the firm. If it was, the second possibility is relevant, while if it was not, the third is the relevant one. Integration must have had some advantage, indirect or direct, for these knowledgeable entrepreneurs with the key to success to have undertaken it. We may thus turn to the other possibilities.

The motive of control was raised in connection with blast furnaces. There it was observed that the larger member of a buying and selling pair often has an incentive to control the smaller. A forge was much smaller than a blast furnace, as Table 5.1 shows, and the output of a single blast furnace could supply several forges. The owner of a forge had little incentive to own a blast furnace to help the operation of his forge, for he could get a standardized supply of pig iron by purchasing from only one furnace if he wanted, while if he bought a blast furnace he would be forced to think of selling that part of the output that he could not use. Rolling mills, however, were more nearly the size of blast furnaces, and rail mills were the largest of rolling mills, by possibly a factor of three or four. It required the output of two or three blast furnaces to supply a rail mill, and the owner of a rail mill could easily want to own these blast furnaces. He would not have to worry about selling their output, as he could use it all, and he would not have to worry about his raw materials, as he could produce it all. In addition, he would have some control over the quality of his inputs.[28] This should not be overestimated, as the mid-nineteenth century was too early for precise quality control to be used, and users of iron often liked to mix their inputs to average out the inevitable differences in quality in the output of any one furnace. This variation will be very important in retarding the growth of economies from integration in Bessemer steel, where the quality differences were very important.

The incentive to integrate for purposes of control can be used to explain the existence of the integration observed. The question remains whether the increased profitability of operation was the crucial factor in making for success of these firms, or whether there was an effect of integration separate

[28] Hunter, "The Influence of the Market . . .," p. 280.

from profitability that made it desirable. The latter is the substance of the last possibility enumerated above, in which the effects of integration independent of its short-range profitability are important. The argument says that changing attitudes toward market domination are an impetus to innovation. The largest firms, or the firms with a structural advantage over other firms, may think that if they can exploit this advantage in conjunction with, let us say, lower costs due to innovation, they could dominate the industry. This would be applicable to the production of rails where there was a high capital requirement and the entry to the field was consequently limited.[29]

There is no doubt that the capital costs of an efficient rail mill were higher than those of a rolling mill making nails or bars or sheet iron. The size of the product was larger and the machinery consequently had to be heavier. The shape of the rail was complex, once edge rails replaced strap rails, and the machinery needed to roll these shapes was not easily made. The need for machines that could deal with difficult problems on a large scale extended to the auxiliary machines as well as to the main rolling trains, making for a very expensive plant.[30] The expensiveness of the initial investment required a large volume of production to spread the fixed cost over many items and keep down the cost of any one. The size of the plant required thus implied both large scale and limited entry, while large scale provided the impetus toward integration already noted.

The problem is similar to the problem connected with the introduction of coke. In both cases there were several interacting changes whose precise pattern of interaction is now known, but is perhaps not important as they were all results of some other, exogenous cause. With coke, the concomitant changes were the growth of coke pig iron production and of demand in the West, and the exogenous movement was the westward growth of the economy as a whole. With the rail mills, we observe output, concentration, and integration all increasing together, and the exogenous cause is the increasingly large-scale machinery of the rail mills.

[29] Richard E. Caves, *Air Transport and its Regulators* (Cambridge: Harvard University Press, 1962), pp. 31-32. Collusion is an alternative path to market domination. See Part II.
[30] *Hunt's, 12* (1845), 67-68; Overman, pp. 358-365.

But while it is reasonable at the present level of aggregation to take the westward growth of the economy as exogenous, it is not obvious that the increasing scale and complexity of rolling mill machinery can be treated similarly. Two possibilities are present. The characteristics of the machinery may have been determined by the nature of the product in demand, while the demand for this product was determined by forces outside the iron industry. Or the attributes of the machinery being discovered and designed may have determined the character of the product. The former possibility asserts that an increase in the demand curve for rails or other heavy iron products brought forward an increased production, while the latter says a shift in the supply curve of rails produced an increased consumption of rails. In the former case, the changes were exogenous to the iron industry; in the latter, they originated there.

The solution to this problem can come only through the identification of the relevant supply and demand curves and an analysis of their changes over time. In the present case the data are fragmentary, and it is doubtful whether theory will be able to discriminate between these hypotheses on the basis of the data for some time to come. The question is therefore left unanswered in this discussion.

The Rail Mills

The more difficult shapes that were being made before 1865 were mostly rails, with structural iron being a smaller category. The rail mills, by definition, were those mills that had adopted the innovations necessary to make this most important type of heavy rolled product. The firms that made structural iron were also rail mills, and the making of heavy products may be attributed exclusively to them. In addition, the major change in rolling mill technique that came from the iron rolling mills, the three-high mill, was the creation of one of the rail mills. It was first introduced at Cambria in 1857 and spread rapidly to other rail mills.

The progress of the rail mills from their inception in the 1840's to the end of the Civil War may be seen as a struggle against prior British domination where the history of technological change was a history of catching up with the advanced

British technology. This does not mean that the solutions to technical problems used in Britain and in America were the same, but only that the American rail mills had to solve certain problems to stay alive, and they concentrated their energies in this direction. The records that have come down to us talk about rail mills rather than rolling mills, and although we may conclude from this and the above discussion of their scale of operations that they led other rolling mills, the extent of their lead cannot be estimated. The problem is similar to that in Chapter 2; rails were certainly important in the shift of technology as well as in the shift of demand, but it is an impossible task to quantify the proportion of the shift that was due to rails, as distinguished from concomitant events.

The location of early rail demand in the East had the important effect of making the suppliers of this demand, whether of the East or of the West, vulnerable to English competition. And when the demand for rails in England fell at the end of the 1840's, and the export price fell as a result, the American ironmasters suffered. The hardships wrought by the English competition were not restricted to the producers of rails, and we have already recorded the attempts of ironmasters to have the tariff raised as a panacea. Cooper and Hewitt summarized the condition of the iron industry for the 1849 tariff convention. Of the rail mills, they said: "Of fifteen rail mills, only two are in operation [December 26, 1849], doing partial work, and that only because their inland position secured them against foreign competition, for the limited orders of neighboring railroads — and when they are executed, not a single rail mill will be at work in the land."

Cooper and Hewitt had an explanation for this state of affairs, with which they prefaced the above statement:

> For the production of bar iron of equal quality from pig iron, we stand on an equality with Wales; in other words, the forge pig iron can be made here at a rate not materially greater than there — except charcoal iron, which generally makes a superior bar iron and commands a better price.
>
> The finished bar iron, however, costs much more to make here than in Wales, and this is chiefly due *now*, to the higher price of *labor* in this country. The difference in the first cost of a ton of railroad iron, we cannot estimate at less than $15. The expense of delivery in New York from well located works in this country, and

from Wales, does not materially differ; — so that, with a duty much less than $15, when, to keep their works in operation, both foreign and domestic makers are willing to sell at cost, our mills are compelled to suspend work. Such is the present state of things — for the duty ranges from $7 to $10, according to the quality of the iron.[31]

This is a typical explanation of the events of the late 1840's by contemporaries. The price of pig iron at this time was about $20 to $25 (for anthracite iron) and the price of iron rails at mills in Pennsylvania was about $50.[32] A price differential of the order of magnitude of $15 was severe enough to sway anyone's purchases, despite the difference in the quality of rails supplied by English and American mills. American iron men had not yet begun to think seriously in terms of quality over price, and the cost differential cannot be attributed to a deliberate attempt by them to raise the quality of rails produced. American ironmasters had technical difficulties, and it is to them that the price differential may be attributed.

The concentration of contemporary observers on the relative price of labor has been commented upon by many later writers.[33] But the problems of the iron industry were not those stemming from this cost. If wages were higher relative to other costs in the United States than they were in Britain, economic theory tells us that American ironmasters should have tried to substitute other factors, such as capital, for labor. This, however, did not happen. English rolling mills used a more advanced technology at this time than the American mills, and this technology appeared to imply a lessened use of labor. Hewitt wrote from England in the 1860's that "the new rolling mills [in England] beat us to death by the use of hydraulic cranes everywhere to lift and carry the iron. They do not employ half the men we do for the same work."[34] If Hewitt's observations were correct, two conclusions are possible. Either the alleged cost differences between America and Britain did not exist, or the technology of ironmaking had developed to a point where labor-intensive techniques were not profitable to use at any existing wage levels. Whichever conclusion is chosen,

[31] Both quotations come from Convention of Iron Masters, *Documents...*, p. 57.
[32] Appendix C, Table C.15.
[33] The most recent investigation to take this as its starting point is the one by Habakkuk referred to in Chapter 1, footnote 7.
[34] Quoted in Nevins, p. 246.

the problems of the American iron industry were not those of reacting to a high relative cost of labor. There was the problem of higher costs in America than in Britain, as has just been shown, but the causes of these higher costs were legion: transportation costs, backward technology, costs of raw materials, and others. It would be an interesting problem to elucidate the precise nature of the international differences of costs,[35] but it is not necessary for this discussion. The development of rolling mill technology in the two countries developed along similar lines, although the specific machines used could vary, and it is possible to talk in terms of the United States "catching up" with Britain in this technology.

This catching-up process began in the production of rails in the 1840's, when American ironmasters entered the field. Many of its technical features have already been described in the discussion of rolling mill technology in general. It remains to show the implications of this process for rail-making in the United States.

In 1845 there were, according to contemporary reports, 4,000 miles of railroads in the United States laid with bar rails, but no interest in this country in the manufacture of T rails.[36] The first American rolling mill to make T rails was at Mount Savage, Maryland, and its example was speedily followed by other mills.[37] The rails at Mount Savage were apparently rolled in 1844, and the years from 1845 to 1848 witnessed the first boom in the construction of American rail mills. The Trenton Iron Company's rail mill was constructed in 1845, that company's initial venture into a new and promising field. In 1846, it was followed by rail mills at Phoenixville, at Danville in Montour County (first called the Montour Iron Works, but changed to the Pennsylvania Iron Works during the Civil War), and by the conversion of the mill at Brady's Bend to the making of T rails.[38]

It is a bad practice to generalize from the timing of events whose dates are not well known or to talk about the first en-

---

[35] As Burn attempts to do for a later period.

[36] *Hunt's, 12* (1845), 67; *ARJ, 18* (1845), 265.

[37] See p. 48.

[38] The chronology is taken from Daddow and Bannan, pp. 694-695. This differs slightly with the account given by Swank, pp. 434-435, which is the source for the conversion of Brady's Bend. Swank includes all the mills he knows about, while Daddow and Bannan only have those that survived till 1865.

trants to a field when we know that the list before us omits firms who tried to enter but were not successful. Nevertheless, the above list is too suggestive to let pass without comment. The five mills listed were all integrated, and they were all firms using mineral fuel. The geographical split, and therefore the split between types of coal, is of great interest. There were three firms from the East, all using anthracite as fuel, and all being among the four largest firms in 1860. There were two firms in the West, and the reader will recall having made their acquaintance in Chapter 3. These firms were two of the three successful users of coke in the blast furnace during the 1840's.

The sample under consideration is too small to provide definitive reasons for its characteristics. The integration may be explained at least partially by the motives of quality control mentioned above, but the geographical distribution shows a confusing pattern. If, however, we exclude the two rather unusual coke firms from our consideration, a locational pattern emerges in the construction of the other rail mills in the 1840's. The mills just listed were followed in 1847 by the Rough and Ready Iron Works at the same location as the Montour Works, the Bay State Iron Works in Boston, and the Lackawanna Rolling Mill in Scranton, Pennsylvania. The last successful mill of this boom was the Safe Harbor Rolling Mill, built in 1848 by the owners of the mill at Phoenixville.[39]

All of these mills used anthracite for fuel, and all were located in the East. All of them, in fact, except the Bay State Works, were located in the anthracite region of eastern Pennsylvania and its watershed. Railroad construction in the 1840's was largely confined to the East, and there was obviously a close connection between these two developments. The manufacture of rails on a competitive basis, and by this is meant competitive with English imports, required the use of the new technology, that is, mineral fuel and the associated innovations in smelting and refining.[40] Although the production of rails was not necessary for the use of the new technology, there was a causal influence working the other way also, and the location of these

[39] Daddow and Bannan, pp. 694-695. This list includes the fuel used by the mills, with the symbols for anthracite and bituminous coal reversed.

[40] This statement needs to be supported by a cost calculation or similar evidence. However, no one appears to have made rails any other way than by using mineral fuel pig iron, puddling, and rolling, and we may use that in lieu of a cost calculation for support of our assertion.

rail mills in the East was both a cause and an effect of the more rapid adoption of mineral fuel in this area. In light of the discussion of Chapter 3, the excess of anthracite pig iron production over rail mill inputs, and the location of the first two rail mills in the West, however, the causal side should not be emphasized over the resultant.

We may say, then, that the location of these rail mills was the result of the available technology and resources and, to a lesser degree, the location of railroad building, that is, of the demand for rails. The technology and known resources were shown to be partially a result of the location of demand, where demand is now the demand for many types of iron. Most rail mills were located in the East in the 1840's ultimately because of the larger demand for all kinds of iron, rails included, in that region.

The depression following the fall in English rail prices at the end of the 1840's stopped the construction of rail mills in the United States and no new rail mills were built before 1852. From 1852 to 1857 there was a second boom in the construction of rail mills, although the lag involved in their construction meant that their production did not arrive soon enough to contribute to the satisfaction of the boom demand that spawned them. The rail mills built in the middle 1840's were all in the East, and, except for the older mills at Brady's Bend and Mount Savage, there were no rail mills west of the Allegheny Mountains. Of the fifteen rail mills built in the middle 1850's, only five were East of the Alleghenies, and the location of the other mills ranged as far West as Chicago and Detroit.[41]

The westward migration of rail mills implied an increasing use of bituminous fuel, and the two increased together. The effect of a ton of rails produced in the West on the demand for pig iron, however, was less than the effect of the same production in the East because rerolling had become more widespread in the former region. Rerolling spread rapidly in the 1850's and was a major factor in reducing the amount of

[41] Daddow and Bannan, pp. 694-695.

[42] *Ibid.* A mill that used bituminous coal was classified as Western. Lesley, p. 763, estimated that almost 150,000 tons of rails were made in 1856, while 100,000 tons of old rails were reworked. The AISA reported that over half the rails made in 1866 and 1867 were rerolled rails also. [*Hunt's*, 58 (1868), 265.] Fogel, in the paper cited, has reworked these and other data in an effort to derive more exact estimates than are used here.

pig iron used to produce a ton of wrought iron. In 1863 and 1864, the volume of rerolled rails exceeded the volume of new rails produced. But while the Eastern rail mills produced more new rails than rerolled ones, the Western mills produced over twice as many tons of rerolled rails as new ones.

The proportion of coke pig iron that was used to produce rails may be roughly estimated. The Western rail mills made 80,000 tons of new rails in the two years 1863-1864, and the production of coke pig iron was 330,000. Allowing a reasonable amount of waste in the conversion of pig iron to rails, the proportion of coke pig iron used for rails was about one-third. This was roughly the same as the proportion of anthracite pig iron used for rails fifteen years earlier, and somewhat larger than the proportion used at this time.[43] Rails were then an important, but not the only, demand for coke pig iron, as they had been for anthracite iron during its early years. The expansion of each was dependent upon the expansion of the other, and both were the results of the westward push of the economy.

Leadership of the iron industry in the decades preceding 1865 was concentrated in the East. This central feature of the "new iron industry" of these years was at its peak during the Civil War, but within the new iron industry were the seeds of its destruction. Demand had been strongest in the East during this period, but the events after 1850 had already altered that condition. The Civil War would perhaps terminate it. Anthracite was the more suitable fuel before the West was fully explored, and the variety of bituminous coal appreciated. But coke was more adaptable to the requirements of technology as it developed after the Civil War, and the raw material base of the Eastern industry, as well as its support from rapidly growing local demand, was superseded by developments in the West. And Western iron ore was also beginning to supplant Eastern deposits.

The Western iron industry, however, did not merely extend the gains of the Eastern industry. Its expansion was linked to the introduction of one of the most important innovations of the post-Civil-War era: the Bessemer converter. The shift of location (involving only a short distance and taking place largely within Pennsylvania and neighboring states) was part

[43] Daddow and Bannan, pp. 694-695; Appendix C, Tables C.2, C.6.

of the transition from the new iron industry of the ante bellum era to the steel industry of later years. The leadership of the iron industry went from the iron plantation to the independent anthracite blast furnace to the rolling mill and the rail mill. The change from iron to steel was the grand achievement of the rail mills and of paramount importance for the economy as a whole; the description of this event and its effects belongs to Part II of this work.

PART II

FROM IRON TO STEEL,
1865-1900

6

New Processes for Making Steel

Bessemer's Discovery

The discovery of the Bessemer process at the middle of the nineteenth century was a critical event in the development of the iron and steel industry and of the railroads as well. This chapter describes the introduction of the Bessemer process and the other methods and improvements in the manufacture of steel that followed it. The following two chapters show the effects of these methods on the firms that used them; the final two chapters of this Part place the making of steel into the context of the iron and steel industry as a whole.

Before Henry Bessemer's famous discovery, steel was made from pig iron in two steps. Carbon was removed from pig iron by puddling to make wrought iron, and then carbon was put back into the iron by reheating wrought iron with charcoal in a steelmaking furnace. This made blister steel, which could be converted into shear steel, a relatively uniform product, by welding several bars together, or into crucible steel, a completely uniform product, by melting pieces in small crucibles. Bessemer set out to make an improved wrought iron for military use. He found that blowing air through molten pig iron removed carbon as cheaply as existing methods and also kept the iron molten. The melting point of iron rises as the carbon is removed, and the difficulties of keeping wrought iron molten had led to the production of nonhomogeneous wrought iron and steel. Bessemer created a uniform product, and he discovered in addition that he could stop the process partway and produce steel. Unfortunately the timing of the stop was critical, and the product was of very uneven quality. Robert Mushet solved this problem by allowing all the carbon to be removed, as in making wrought iron, and then adding some back as in the older processes of steelmaking. All this was done while the metal was molten, and the product had the uniform consistency of crucible steel, although it lacked the high quality of that product.[1]

A similar discovery was made in this country about the same time by an ironmaster named William Kelly. He did not develop the process to the point of commercial usefulness, but he developed it far enough to be able to convince the patent commissioners that he should get a patent in preference to Bessemer for the process.[2] The parallel discovery of this process by two people working independently may indicate that there was some kind of "pressure" for the discovery of a new process in the iron and steel industry, but the main importance of the parallel discovery for us is its creation of conflicting patents in the United States. Kelly had the process patent, Bessemer had patents for the machinery, and Mushet had a patent for the replacement of the carbon. All these patents were in existence by 1857, and although the English rights to Mushet's patent were soon lost by a tax default, the American rights continued in force.

The first public account of Bessemer's invention was in the paper he delivered to the British Association for the Advancement of Science in August, 1856. Although many people were skeptical of Bessemer's claim, the importance of the paper, if true, was widely recognized, and it was widely reported in America.[3] Abram Hewitt was among the first to react to the paper, and he built an experimental Bessemer converter at Trenton soon after he learned of Bessemer's paper. The converter, however, was never used; the people at Trenton heard about the "failure" of the Bessemer process and abandoned their project.[4]

There were two failures of the initial Bessemer process. The first was the inability to yield a uniform carbon content,

[1] Sir Henry Bessemer, *An Autobiography* (London: Offices of "Engineering," 1905).

[2] John Newton Boucher, *William Kelly: A True History of the So-Called Bessemer Process* (Greensburg, Pennsylvania: Published by the Author, 1924); James M. Swank, *History of the Manufacture of Iron in All Ages* (second edition; Philadelphia, 1892), pp. 396-400.

[3] *Hunt's*, 35 (1856), 499-500; *ARJ*, 29 (1856), 595; *JFI*, 3rd series, 32 (1856), 267.

[4] W. F. Durfee, "The Manufacture of Steel," *Popular Science Monthly*, 39 (October, 1891), 729-749.

which was solved by Mushet's innovation. The second was that this process, unlike puddling, did not remove phosphorus, and that therefore only a restricted class of iron could be used. It took time to discover this fact and to find a suitable class of iron. The knowledge of chemistry was sufficiently rudimentary at that time, in addition, that the only way of discovering if an iron was suitable was to try it. As late as 1866, the owners of the American patent rights could only repeat Bessemer's claim that iron containing more than 0.1 per cent sulphur, 0.075 per cent phosphorus, or 1.75 per cent silicum was not *"well* adapted" to the Bessemer process, and state: "We are prepared to test irons at seventy-five dollars per ton of pig, the iron to be delivered, and the product to be removed at the owner's cost. This test includes sufficient hammering, rolling, and cold bending to determine the various practical qualities of the product."[5]

The date of this quote, 1866, is usually taken as the start of the commercial manufacture of Bessemer steel in America. The decade between the publication of Bessemer's paper and the adoption of his innovation in America was a time when many problems were solved, or at least made manageable. The problems may be grouped into three classes: legal, technical, and financial. The legal problems were the result of the conflicting patents, about which much unrecorded bargaining took place. The second group of problems includes the chemical problems just mentioned and other defects of the early Bessemer manufacture. We know the technical solutions adopted, but little about the search for them. The third class of problems are those that derived from the difficulty of convincing people to use the new process. Without an adequate knowledge of the technical and legal problems, it is hard to say much about the financial problems of these years.

The legal chronology can best be summarized by an extended quote from Swank's history of the industry.

As early as 1861 Captain E. B. Ward, of Detroit, and Z. S. Durfee, of New Bedford, obtained control of the patents of William Kelly, and in that year Mr. Durfee went to Europe to study the Bessemer process. During his absence Captain Ward invited Mr. William F. Durfee, also of New Bedford and a cousin of Z. S. Durfee, to erect an experimental plant at Wyandotte, Michigan, for the manufacture

[5] *Bulletin, 1* (December 5, 1866), 97.

of pneumatic [Bessemer] steel, and this work was undertaken in the latter half of 1862.

In May, 1863, Daniel J. Morrell, of Johnstown, and William M. Lyon and James Park, Jr., of Pittsburgh, having become partners of Captain Ward and Z. S. Durfee in the control of the patents of Mr. Kelly, the Kelly Pneumatic Process Company was organized, Mr. Kelly retaining an interest in any profits which might accrue to the company. It was resolved to complete the experimental works already undertaken, and also to acquire the patent in the country of Mr. Mushet for the use of spiegeleisen as a recarburizing agent. The patent was granted in England in 1856 and in this country in 1857. Mr. Z. S. Durfee accordingly went to England to procure an assignment of Mr. Mushet's patent. The latter purpose was effected on the 24th of October, 1864, upon terms which admitted Mr. Mushet, Thomas D. Clare, and John N. Brown, of England, to membership in the Kelly Process Company. On the 5th of September, 1865, the company was further enlarged by the admission to membership of Charles P. Chouteau, James Harrison, and Felix Valle, all of St. Louis. In September, 1864, William F. Durfee succeeded in making Bessemer steel at the experimental works at Wyandotte. *This was the first Bessemer steel made in the United States.* A part of the machinery used at the Wyandotte works was an infringement upon Mr. Bessemer's American patents.

The control in this country of Mr. Bessemer's patents was obtained in 1864 by John F. Winslow, John A. Griswold, and Alexander L. Holley, all of Troy, New York, Mr. Holley visiting England in 1863 in the interest of himself and his associates. In February, 1865, Mr. Holley was successful at Troy in producing Bessemer steel at experimental works which he had constructed for his company at that place in 1864. Mr. Mushet's method of recarburizing melted iron in the converter was used at Troy, and this was an infringement upon his patent in this country.

As the Kelly Process Company could not achieve success without Mr. Bessemer's machinery, and as the owners of the right to use this machinery could not make steel without Mr. Mushet's improvement, an arrangement was made by which all of the American patents were consolidated early in 1866. Under this arrangement the titles to the Kelly, Bessemer, and Mushet patents were vested in Messrs. Winslow, Griswold, and Morrell, the first two being owners of seven-tenths of the property and Mr. Morrell holding the other three-tenths in trust for the Kelly Process Company. This arrangement continued until the formation of the Pneumatic Steel Association, a joint-stock company organized under the laws of New York, in which the ownership of the consolidated patents was vested. Z. S. Durfee acted as the secretary and treasurer of the company. The ownership of the patents was afterwards vested in the Bessemer Steel Company Limited, an association organized in 1877 under the laws of Pennsylvania. This association has been succeeded by the Steel Patents Company, organized in 1890. All the original English and American patents have expired. The Steel Patents Company, however, owns other patents relating to the manufacture of Bessemer steel which have not expired.[6]

This account completely ignores the technical and financial problems involved. One aspect of the technical arrangements which resulted from the legal arrangements was the separation of investigators into two distinct laboratories, one under the direction of Holley and the other under Ward and Durfee. Holley's work determined the future form of Bessemer steel mills, despite Durfee's earlier success at making steel, which may have determined the seventy-thirty split of patent ownership in 1866. But we do not know, and it is likely that financial considerations were more important.

The expanding ownership of the Wyandotte company suggests that Ward and Durfee were in some kind of financial difficulty and needed frequent infusions of capital to continue their investigations. This contrasts with the stable membership of the Troy group and may be an index of the relative financial power of the two groups at the time of their merger in 1866. A motive for expansion of the part of Ward and Durfee was to get people to use their patents. They admitted four parties into their company in 1863, according to a report somewhat at variance with Swank's, with promises to use the new process. Only one ever did, and that was not until after a delay of seven years, too late to have any influence on the patent merger of 1866.[7] The Troy group, on the other hand, had found someone who was willing to use the new process, without diluting the ownership of the firm. The aspiring steelmaker was the Pennsylvania Steel Company, a subsidiary of the Pennsylvania Railroad, which owned one-third of its stock. There were some complicated financial dealings in this transaction, which was initiated in 1865, but they were apparently related

[6] Swank, pp. 409-410.
[7] *Bulletin, 30* (September 1, 1896), 195.

to debts of Holley and do not indicate financial difficulty on the part of Winslow and Griswold.[8]

The fragmentary nature of the data that have come to light force us to leave as open questions the severity of the technical problems of these years, the nature of the financial problems that seem to have beset these two groups, and the extent of their legal conflict. We know only that they had solved many of the technical problems implicit in the manufacture of Bessemer steel by 1866, and that at that time they decided to merge their interests rather than to battle each other in the courts.[9] As indicated above, the terms of the merger may be made to sound reasonable, but all such reconstructions are highly speculative in the absence of additional data.

An incentive for the two Bessemer groups to solve their problems was provided by the demand for steel rails, which was rising rapidly just after the Civil War due both to the increasing dissatisfaction of railroad men with iron rails and to the educational work done by salesmen of the British steel producers. "Full credit must be given to the English steelmakers for creating a market for steel rails by fairly forcing them on railroads. Practically the whole of the pioneer educational work among American railroad men was done by English drummers."[10]

The increase in the weight and speed of locomotives had effects on iron T rails which were becoming troublesome by the later 1850's. Some of the dissatisfaction with iron rails has been noted in Chapter 5; the expensiveness of frequent rail changes deriving from labor charges and from the delay to traffic was well known. This dissatisfaction was increased by

[8] Agreement to modify the license granted by Winslow and Griswold to the Pennsylvania Steel Company in accordance with the transfer of ownership of patents in return for $1 paid to Morrell, April 14, 1866. Letter from Blatchford to Winslow, January 22, 1867. (These manuscript documents were furnished to me by Elting E. Morison. They are part of the business records of Blatchford, Seward, and Griswold, Attorneys, 29 Nassau Street, New York, and their successor firm.) George H. Burgess and Miles C. Kennedy, *Centennial History of the Pennsylvania Railroad Company* (Philadelphia: The Pennsylvania Railroad Co., 1949), p. 293.

[9] *Bulletin, 12* (May 15, 1878), 116; Durfee, "The Manufacture of Steel, continued," *Popular Science Monthly, 40* (November, 1891), 15-40.

[10] Herbert N. Casson, *The Romance of Steel* (New York: A. S. Barnes and Co., 1907), p. 24.

the demands of the Civil War for rapid service with a minimum of maintenance, and during the war railroads began to experiment with steel rails. The Pennsylvania Railroad discussed the rail question in its annual report for 1863, noted that European railroads had started to use steel and ordered 150 tons of steel rails for experimental purposes. The annual report for 1864 says only that "rails became constantly a more difficult problem," but in 1866 further imports of steel rails were made. The experiment started in 1863 was considered a success, and steel rails were held economical despite their initial cost of double the price of iron rails.[11] The sequence of events was typical; by the end of 1866 the British steel firm of Messrs. Charles Cammell and Company, Ltd., Sheffield, had orders for Bessemer steel rails from several American railroads, including the Erie, New York Central, Pennsylvania Central, Boston and Worcester, and Boston and Providence.[12]

Hewitt attempted to fill some of the demand for high-quality rails by producing a steel-headed rail, using a combination of wrought iron and the older types of steel.[13] Once the initial problems of the Bessemer process were dealt with, however, the Bessemer process became by far the cheapest way to satisfy this demand, and Hewitt's steel-headed rails were abandoned. It took 7 tons of coal to make 1 ton of cast (crucible) steel from pig iron by the old method: $2\frac{1}{2}$ tons of coal to make blister steel and $2\frac{1}{2}$ tons of *coke* to make cast steel. The same process with a Bessemer converter required $\frac{2}{10}$ of a ton of coal to heat the converter plus about $\frac{6}{10}$ of a ton to melt the iron. One part fuel in the Bessemer process equaled 6 or 7 parts in the old method of steelmaking, and a comparison of labor and machinery requirements would yield similar results.[14] A comparison with wrought iron was more usual, and the steel producers asserted that the cost of making Bessemer steel was only slightly greater than the cost of making puddled iron.[15] Given the relative

[11] Burgess and Kennedy, pp. 292-293.

[12] *JFI*, 3rd series, *83* (January, 1867), 22; *Bulletin, 1* (December 5, 1866), 97.

[13] Allan Nevins, *Abram S. Hewitt* (New York: Harper and Brothers, 1935), p. 235.

[14] Samuel Harries Daddow and Benjamin Bannan, *Coal, Iron, and Oil, or the Practical American Miner* (Pottsville, Pennsylvania, 1866), p. 647.

[15] Alexander L. Holley, *The Bessemer Process and Works in the United States* (from the "Troy Daily Times," July 27, 1868; New York, 1868), p. 7.

merits of the two metals, Bessemer steel was sure to capture the expanding market for rails, even if the high quality of crucible steel would ensure it of a market for other products.

The Bessemer Process In Use

The owners of the combined patents lost no time in trying to attract people to use their patents. They placed advertisements for licensees in the AISA *Bulletin* on a continuing basis starting at the beginning of 1867. They stated that the patents for "the Pneumatic or Bessemer Process" had been consolidated under a trusteeship, and named Z. S. Durfee as the agent to whom one could apply for a license.[16] They then issued a pamphlet giving more information on the costs of the new process.[17] The cost of a plant with 2 three-ton converters was given as $80,000; of a "five-ton plant" with steam power, $125,000; and of first-class apparatus with fireproof buildings and duplicate machinery making 50 tons of ingots in twenty-four hours, $200,000. It cost only two-thirds as much for a Bessemer plant as for a crucible steel, charcoal bloom, or puddled bar plant of the same capacity, the trustees asserted, and it took only 30 men to run a five-ton Bessemer plant.

Royalties for the new process were set in sterling to harmonize with Bessemer's charges in England. The base charge was one pound sterling for each gross ton of iron used to make ingots for rails, and higher rates were charged for ingots for other purposes. Making an allowance for the waste in conversion, the charge was about $5 per ton for the use of the Bessemer process. It was asserted that the agreement of 1866 greatly lowered the royalties charged,[18] but evidence is lacking to support this contention, and it may be only a myth of the steel industry.

In addition there was an initial cost to a licensee of $5,000, in return for which Winslow, Griswold, and Morrell furnished plans of a plant and information on the processes involved. The accounts of the licensee firm had to be kept open to Win-

[16] *Bulletin,* 1 (February 27, 1867), 212.

[17] John F. Winslow, John A. Griswold, and Dan'l J. Morrell, Trustees, and Z. S. Durfee, General Agent, *The Pneumatic or Bessemer Process of Making Iron and Steel* (Philadelphia, 1868).

[18] *Iron Age* (May 9, 1878), pp. 15-16.

slow, Griswold, and Morrell in order to provide for checks on royalty payments. The works and processes had to be open to Winslow, Griswold, and Morrell also, and Winslow, Griswold, and Holley opened the Troy works to licensees in return. They would employ two people at a time from a licensee firm to work at Troy in the first two years of the license, although they would not pay their wages. In contrast, however, with this free communication among licensees and licenser, none of the information supplied by Winslow, Griswold, and Morrell was to be communicated to anyone not a licensee.[19]

These provisions for communication are the most interesting feature of the licenses, and they raise the question of whether the institutional form adopted by an industry affects its technological development. The communication facilities embodied in the licenses were widely used, and new ones were added. Almost all the early Bessemer works were built according to plans drawn by Holley, presumably acting as part of the licensing firm. Of the eleven plants in operation in 1880, Holley designed six, consulted on the construction of three more, and was the inspiration for the remaining two which were copied after one of the first six. The technical personnel of these firms were involved in a continuing game of musical chairs, and the managers of the newer works usually had been trained at one of the earlier ones.[20] Frequent meetings were held of the five or six top engineers of the industry to discuss common problems. The principal participants of these meetings were the following: Holley; John Fritz, who was then at the Bethlehem Iron Company where the meetings were often held; George Fritz, John's brother, who had taken over for him at Cambria; Captain R. W. Hunt, the author of the industry history just cited and at that time the manager of the Troy works; and Captain William Jones, the brilliant manager of the Edgar Thomson Steel Works, Carnegie's plant.[21]

[19] License for Kelly, Bessemer and Mushet patents (n.p., n.d.). The license also provided that the licensee would commence work within a year and that the agreement applied to all future patents of the three men named owned by the licensing parties.

[20] Robert W. Hunt, "A History of the Bessemer Manufacture in America," AIME, 5 (June, 1876), 201-216.

[21] John Fritz, The Autobiography of John Fritz (New York: John Wiley and Sons, Inc., 1912), p. 160; E. B. Coxe, another engineer and author of the industry, often attended these meetings also. George Fritz died in 1873 [E & MJ, 16 (1873), 153].

Finally, Holley wrote a series of confidential reports on technical subjects which were distributed among this circle. On their title page they bore the following legend: "These papers are printed, not as a publication, but for the convenience of my clients, and for their exclusive use." The reports were only a few pages in length, and they each treated a specific subject: a noteworthy feature of a plant, a new process, a new machine. They were published in two series in the years 1874-1877, and there were fifteen to twenty of them.[22] After the formation of the Bessemer Association (that is, the Bessemer Steel Company) in 1877, they were for the exclusive use of its members. Holley is said to have regarded them as his best work, and his clients found them extremely useful.[23] The uniform size of the Bessemer steelworks in existence in 1880 has been attributed to the newness of the process and the "immaturity of firms,"[24] but it would seem that the extensive communication among firms, the all-pervading influence of Holley, and the almost exclusive production of one product—rails—would account for this phenomenon.[25]

[22] The following reports are in the possession of the Library of Congress:
First Series (1874-1875) and Supplements (1875-1876)
  3   The Siemens steel manufacture, from pig and ore, at Landore and at the steel works of Scotland.
  4   West Cumberland iron and steel works.
  5   The Pernot revolving hearth furnace and practice, for Siemens-Martin steel and for puddling.
  6   Brown, Bayley and Dixon's steel works.
  7   Barrow haemetite iron and steel works.
  8   The steel manufacture in France and Belgium.
Suppl. to   2   The value of manganese. The practice in Germany.
2nd Suppl. to   2   The effect of manganese, carbon, phosphorus, and other ingredients on iron and steel.
Second Series (1877)
  2   The direct use of the blast-furnace metal in the Bessemer process.
  3   Price's retort furnace for reheating and puddling.
  4   Report on the phosphorus steel manufacture, with and without the Sherman process.
  5   Rolling mill improvements. Rolling double-length rails direct from the ingot, blooming-tables for reversing trains, etc.
  6   The hot-blast cupola and utilizing converter flame for heating cupola blast.
  7   Solid steel castings for ordnance, structures, and general machinery, by the Terrenoire process.
[23] American Institute of Mining Engineers (AIME), *Memorial of Alexander Lyman Holley* (New York, 1884), p. 136.
[24] R. N. Grosse, "Determinants of the Size of Iron and Steel firms in the United States, 1820-1880," unpublished doctoral dissertation, Harvard University, 1948, p. 214.
[25] For the proportion of Bessemer steel used to make rails, see Appendix C, Table C.11.

The communication network established in the early Bessemer steel industry appears to have been designed in large part to make use of the extraordinary talents possessed by Holley. The success of these arrangements is evident in the great influence of Holley and the extent to which his innovations helped the American industry. R. W. Hunt stated in the memorial to Holley, "An imperfect knowledge of the chemical requirements of the [Bessemer] process, an utter absence of tested and approved refractory materials, and, above all, imperfect machinery, were the conditions of the problem which, in 1864, Holley set himself to solve."[26] He worked in all these areas, and his reports and inventions reflect his wide range of interests.[27] But his primary importance was probably in the last-named category, the improvements in machinery. Two major innovations may be noted: the "American" or Holley floor plan, and the Holley bottom.

The original British floor plan for a Bessemer plant had two converters facing each other across a deep pit which contained molds to receive the molten metal from the converters. Two converters were used because the operations of a converter were discontinuous, and the output of a single converter would occupy the auxiliary equipment only a short part of the time. Heavy auxiliary equipment was needed to power the Bessemer machinery itself, and also to move and process the large amounts of metal the Bessemer converter used. The indivisibility of this machinery meant that large plants were necessary for the efficient utilization of the Bessemer process and that large outputs were needed for the success of these plants. The converters only worked a small percentage of the time because of the technical characteristics of the production, the difficulty of moving the metal from place to place, and the time needed to rebuild the converter lining consumed in the extraordinarily high heat of the process. The bottom of the converter would wear out after one to three heats, the converter would have to be cooled, and a man would climb inside and repair the lining of the converter.[28] Having two converters kept the machines employed more of the time, but the original plan was not satisfactory.

[26] AIME, *Memorial* . . ., p. 31.

[27] For a list of Holley's inventions, see *ibid.*, pp. 69-70; *Bulletin, 16* (November 15 and 22, 1882), 309.

[28] Fritz, pp. 155-156.

Holley set himself to improve the speed with which the converters could be used by two means. He arranged the converters to facilitate the movement of the metal, and he shortened considerably the time required to repair the converter lining. The first was done by placing the two converters side by side instead of facing, raised high off the ground to let them discharge their contents at ground level rather than in a pit. This arrangement opened up the working space around the converters and permitted further innovations to speed the handling of materials.[29] But the converter lining still wore out with great regularity and slowed operations behind the speed of the auxiliary equipment. Holley introduced his new bottom in 1869-1870, and received a patent on it in 1872.[30] The previous practice had been to repair the lining inside the converter shell, which meant cooling it sufficiently for a man to get inside. Holley's innovation was to make the bottom of the shell removable. A worn-out bottom could be taken off the converter and a new one put on without cooling the converter itself; the saving in time is obvious. Holley claimed for himself the use of refractory brick, the preformed bottom, and the predrying of the bottom. The Holley plan gave the converters a "front" where the metal could be handled with facility; it also gave them a "back" where the bottoms could be handled equally expeditiously.

These innovations were far from the only ones introduced at this time, but they were among the most important, and they characterize the spirit of the innovations. Holley and Fritz are the major names in this process, but there were many active engineers whose influence cannot be accurately estimated from this distance. A race among steel men developed as each tried to increase the speed of his plant. Holley's innovations started and permitted this race, and many subsequent improvements continued it. The record of increasing speed could be described in detail, but without an explicit idea of the cost structure of steel mills, the proliferation of numbers

[29] The new layout was patented as "a combination of crane, converter, and chimney." AIME, *Memorial . . .*, p. 69. Pictures of a Bessemer plant around 1890, with the Holley floor plan, are given in Carnegie Brothers and Company, Ltd., *The Edgar Thomson Steel Works and Blast Furnaces* (Pittsburgh, 1890).

[30] U.S. Patent 133,938, granted December 17, 1872.

would serve little function.[31] In 1878 the twenty British Bessemer works in operation produced 800,000 tons of steel.[32]
In the same year, the ten active American firms made 650,000
tons,[33] about half again as much apiece as their British counterparts. By 1880, the American firms had individual capacities
of over 100,000 tons, and production was not far under capacity.[34]

We would like to know if the ever-increasing size of the
American steelworks was economical. The great emphasis
placed upon it indicates that it was, but it does not provide
a test. The similarity of American steelworks likewise precludes
a firm test. Information has survived about a few of the unsuccessful aspirants to membership in the Bessemer industry, but
usually without enough information to know why they failed.
The National Iron Armor Company of Chester, Pennsylvania, is
typical of these firms. It obtained a license for the Bessemer
process before 1868 and built a plant with five-ton converters.
Holley commented that the works were well-located and well-
constructed. But we hear no more of them, and the causes of
their failure are unknown.[35]

The Freedom Iron and Steel Works of Lewistown, Pennsylvania, is the only early and unsuccessful entrant to the industry about which information has survived. It made its first
"blow" on May 1, 1868, and failed in 1869. The Bessemer works
were dismantled and taken to Joliet, Illinois, where they were
used by the Joliet Steel Company.[36] The causes for this failure
were apparently manifold, for all of the obvious things that
could be wrong with a steelworks were ascribed to the Freedom
Works. First, it was the only American steel mill to use the
English floor plan in preference to Holley's.[37] In addition, the
iron used contained too much phosphorus, and the product

[31] Clark documents part of this race: Victor S. Clark, *History of Manufactures
in the United States* (3 vols.; New York: McGraw-Hill Book Company, Inc.,
1929), Vol. II, pp. 264-265.

[32] J. C. Carr and W. Taplin, *History of the British Steel Industry* (Cambridge:
Harvard University Press, 1962), pp. 97, 108.

[33] Appendix C, Table C.4.

[34] Grosse, pp. 281-282; Appendix C, Table C.4.

[35] Holley, *The Bessemer Process* . . ., pp. 12, 37; Winslow, Griswold, and
Morrell, p. 30. Other obscure aspirants were listed in National Association of
Iron Manufacturers, *Statistical Report for 1872* (Philadelphia, 1873), p. 132.

[36] Swank, p. 411; National Association of Iron Manufacturers, p. 21.

[37] Holley, *The Bessemer Process* . . ., p. 35.

was of bad quality.[38] Finally, the Freedom Company was probably undercapitalized due to the failure of its president to appreciate the large scale of the new process.[39] Each of these factors would have been sufficient to cause failure, and the effects of each of them alone cannot be estimated.

The similarity of Bessemer steel plants at 1880 and our ignorance of the precise causes of the failures in the Bessemer steel industry before that date, then, preclude direct testing of the assertion that high speeds were necessary for these plants. Nevertheless, the indivisibility of Bessemer machinery must have been a potent force leading to faster operations, and the participation of all Bessemer steel plants in the developments leading to greater speed implies that they offered benefits by reducing costs.

More Discoveries

Alexander Holley died in 1882. His passing marked, and may have caused in part, a shift in the development of the steel industry and a decline in the uniformity of steel plants. Perhaps the most important of the changes in the supply of steel after 1880 was the rise in the importance of the open-hearth process. This competitor to the Bessemer process was invented a decade after Bessemer's discovery, but it became important in America only after an additional delay of about two decades.[40] We must understand why the Bessemer process was more suitable for use in the 1870's and 1880's and why the open-hearth process was used in increasing proportion after those decades before inquiring further into other changes in steelmaking. To do this, we must describe the open-hearth process and an important innovation of the 1880's; the basic process.

The open-hearth process may be seen, in retrospect, as a

[38] Durfee, "The Manufacture of Steel, continued," p. 23; *Iron Age* (April 18, 1878), p. 5.

[39] Andrew Carnegie, *Autobiography of Andrew Carnegie* (Boston: Houghton Mifflin, 1920), p. 178. Carnegie may have been connected with the Freedom Iron Company, the parent company of the Freedom Iron and Steel Works, in which case he would be in a position to know about capital supplies. In his statement of income for 1863, he recorded $250 from the Freedom Iron Company. Carnegie MSS, Manuscript Division, Library of Congress, Vol. III.

[40] Appendix C, Table C.5.

logical development of the puddling process. The puddling process reduced the amount of labor necessary to remove the carbon from pig iron by increasing the heat to which the iron was subjected. The open-hearth process increased the heat still further and, by exceeding the melting point of wrought iron, was able to dispense entirely with the need for manipulation of the iron. The iron was placed on a hearth open to the flames from the fuel being burnt, much as in a puddling furnace. The difference was that the fuel, gas, was burnt in the regenerative stoves introduced by William Siemens. In these stoves, the gas was alternately burned in one of two compartments of firebrick checkerwork while the exhaust from the hearth was drawn through the other. The combustion chamber was thus preheated by the exhaust fumes. Heat from the burning gas was added to this existing reservoir of heat, and new heights of temperature were attained.

The open-hearth furnace, in other words, was a puddling furnace to which greater heat could be supplied and, consequently, in which there was no need for a puddler. The contrast between this new process, first introduced in 1866, and the Bessemer process is apparent. The latter process was based on a new source of energy, the impurities in the iron itself, and it required for its operation a set of machinery unlike any that had existed before. Both processes, however, employed high heat to eliminate labor, and they both produced a uniform product that was immediately labeled steel.[41]

The first people to try the open-hearth process were, of course, Cooper and Hewitt. They built an open-hearth furnace at Trenton in 1868, but it was not successful. Present as an assistant at this experiment, however, was Samuel T. Wellman, who played a role in the introduction of the open-hearth process similar to that played by John Fritz in the rail mill and Holley in the Bessemer steel mill. He was present at two of the four open-hearth furnaces built in the following five years, and he was able to correct the errors in design that had led to failure at Trenton.[42]

[41] See AIME, *Memorial . . .*, pp. 171-181, and references given there.

[42] Joseph G. Butler, Jr., *Fifty Years of Iron and Steel* (fourth edition; Cleveland: The Penton Press Co., 1923), pp. 68, 180; *The Otis Steel Company — Pioneer, Cleveland, Ohio* (Cambridge, Massachusetts, privately printed, 1929), no page numbers.

Wellman was not as towering a figure as Holley or Fritz, either because of his native ability or because they were still active. Holley saw the possibilities of the open-hearth process while he was still engaged in improving the Bessemer manufacture. His reports to his clients talk about the newer process, he worked with Wellman at some of the early open-hearth furnaces, and he was connected with the introduction of the Pernot revolving open-hearth furnace. This was one of the major innovations of the period, although it is unclear how much effect it had.[43]

The major problem of the early open-hearth furnaces was maintenance of the furnace in the face of the high heat attained.[44] This was similar to the problem of the Bessemer converter lining, but the solution adopted was not clear. It would appear that the development of better refractory materials was the only major change in this area.[45] Another difficulty of the process was the cost of labor required to charge the furnace. The open-hearth furnace, in contrast with the Bessemer converter, did not start out using molten inputs and could not be charged in the easy way that Bessemer's machinery allowed. Wellman introduced a machine that charged open-hearth furnaces, saving not only labor but the capital costs implicit in cooling the furnace for hand charging. They were universally adopted and became a standard feature of open-hearth works.[46]

The open-hearth process was not widely used until after 1885, due to the high cost of steel made this way before that time, but there were reasons for a few people to adopt it. One reason, of course, was the quality of the product. This may have been real or fancied at this early date, but some people thought the quality of open-hearth steel was more controlled.[47]

[43] AIME, *Memorial* . . ., p. 32; *Bulletin*, 13 (June 4, 1879), 138; Harry Huse Campbell, *The Manufacture and Properties of Iron and Steel*, (second edition; New York: The Engineering and Mining Journal, 1903), p. 206.

[44] *Bulletin*, 11 (December 26, 1877), 337.

[45] Note the existence of a furnace "rebuilding fund" for the open-hearth process in U.S. Commissioner of Corporations, *Report on the Steel Industry* (Washington, 1913), Vol. III, p. 166.

[46] Frank Popplewell, *Some Modern Conditions and Recent Developments in Iron and Steel Production in America* (Manchester: The University Press, 1906), pp. 95-96.

[47] *Iron Age* (April 27, 1876), p. 11; Clark, Vol. II, p. 267. Clark (pp. 270-275) appears to accept the Bessemer steel men's reasoning that there was no important difference between the two types of steel at any time.

Bessemer steelworks also had the advantage of plentiful scrap, and some mills adopted this process to use the scrap.[48] Non-Bessemer plants had the inducement, conversely, that the rights to use the Bessemer process were not available for part of this period. The small size of the open-hearth plant, present for all users, was an additional factor inviting people to use it. The minimum size of an efficient open-hearth plant was far smaller than that of a Bessemer plant, the average sizes (by capacity) of the two types of plant differing by a factor of ten in 1880, although innovations such as machine charging were narrowing the gap after that date.[49] The difference followed from the relative time needed to transform iron into steel in a Bessemer converter (about twenty minutes) and in an open-hearth furnace (six hours or more) and the cost of capital per unit of output implied by this at different outputs. The more "divisible" process was adopted in the interstices of the economy in the 1870's, there to wait until the discovery of the basic process which would lower its costs.

Both the Bessemer and open-hearth processes were initially introduced with an acid lining in the refining vessel, the so-called acid process. This variant was the one in use in 1880, and it is the costs of the two acid processes that were referred to in the immediately preceding pages. In 1879, two Englishmen, Thomas and Gilchrist, introduced a practical way for using a basic lining in making steel and introduced the "basic" process.[50]

The problem with the acid process was that it did not remove phosphorus in the course of refining the iron; the phosphorus in the iron would not combine with the acid lining. As phosphorus was injurious to the metal being produced, the acid process could only use iron made from ore that did not contain much phosphorus. A whole class of iron ores was therefore excluded from use with these processes, as has been mentioned above. The basic process substituted a basic lining for an acid one, allowing the phosphorus in the iron to combine with

[48] *Iron Age* (April 13, 1876), p. 1; American Iron and Steel Association, *Directory to the Iron and Steel Works of the United States* (Philadelphia, 1876-1900), 1876-1880.

[49] Grosse, pp. 281-282; U.S. Census of Manufactures, 1905, Vol. IV, *Special Reports on Selected Industries* (Washington, 1908), 14.

[50] Carr and Taplin, pp. 98-100; Lillian Gilchrist Thompson, *Sidney Gilchrist Thomas: An Invention and Its Consequences* (London: Faber and Faber, 1940).

the lining and be carried off in the slag. The use of a basic lining changed the cost structure of the steel industry, and it was its invention, more than any other single factor, which shifted the balance between Bessemer and open-hearth steel.

The costs of producing steel in the nineteenth century, however, are hard to discover. From the moment the Bessemer process was adopted to any large extent, the steel men came under fire for the supposed effects of their restrictive policies and the protective tariff they worked so hard to get.[51] They retaliated by telling folk tales about what was happening in the industry, leaving the investigator to guess about their costs and profits. The best cost comparison that we have, therefore, is not as good as could be desired, but its explicitness gives us as much information as could be hoped.

The costs are the fruits of an investigation of the steel industry undertaken by the United States Commissioner of Corporations after 1900. The investigation was designed to show the effects of the formation of the United States Steel Corporation, but if the data can be trusted, they may communicate other information as well.[52] The costs given are all for the period after the turn of the century, and they therefore show the structure of costs after the steel industry had shifted its direction toward open-hearth steel. We may examine these costs and attempt to infer from them what the structure of costs was before the decision was made. Some of the relevant data are given in Table 6.1.

The costs of the two processes were very close, and the decision to use one rather than the other would appear not to have been dominated by cost savings. Nevertheless, there was nothing to be lost by using the basic open-hearth process at this point, and even a small something to be gained. This structure of costs made steel men receptive to arguments in favor of this process, and we ask how it came about.

The last column of Table 6.1 shows that the small difference in the works costs of the two methods was the result of two offsetting influences. The works costs were the costs connected with the actual steelworks, as opposed to those costs that were

[51] See Chapter 8.
[52] U.S. Commissioner of Corporations, *Report on the Steel Industry* (Washington, 1913).

TABLE 6.1
AVERAGE COSTS OF STEEL INGOTS
IN THE PITTSBURGH DISTRICT,
1902-1906
($ per gross ton)

	(Acid) Bessemer (1)	Basic Open-Hearth (2)	Difference (2) − (1)
Materials	15.20	14.12	−1.08
Other works costs	1.34	2.13	+0.79
Total works cost	16.54	16.25	−0.29
General and miscellaneous expenses	0.83	0.69	−0.14
Total book cost	17.37	16.94	−0.43

Source: U.S. Commissioner of Corporations, III, 166. The Bessemer billet ingot figures were used in preference to the rail ingot figures.

part of the entire firm and could not be allocated directly to any single plant. Works costs can be broken down, as is done in the table, into material and nonmaterial costs. The former consisted primarily of the cost of pig iron and scrap, less than $.50 being charged to manganese and limestone. The latter consisted of a variety of small expenses of which the most easily recognizable are perhaps labor, fuel, steam, supplies and tools, and materials in repairs and maintenance. The table shows that the nonmaterial costs for the basic open-hearth process were higher than those of the acid Bessemer, while the material costs were lower.

These two offsetting influences may be attributed to different causes. The nonmaterial costs of the acid open-hearth process were as high or higher than those for the basic open-hearth process, not only for this late date, but also earlier.[53] As far as we know, also, they were both higher than the equivalent costs for the Bessemer process all through this period, and we may say that the higher nonmaterial costs of the newer process were due to the fact that it used an open-hearth and not that it used a basic lining.

Material costs are a different story. The acid open-hearth and the acid Bessemer process used the same kind of pig iron,

[53] Ibid., III, 439; Iron Age, 50 (July 21, 1892), 105.

and there is no evidence that the use of scrap reduced the material costs of the acid open-hearth process below the Bessemer costs. The difference here was the result of the difference in lining which enabled a different class of iron to be used, the pig iron suitable for the basic process being cheaper than the pig iron suitable for the acid process.

The change from the Bessemer to the open-hearth process, leaving all other things unchanged, therefore, raised costs, and a change from the acid to the basic process, all other things again being constant, lowered them. These two influences offset each other, and the change from the acid Bessemer to the basic open-hearth process had little effect on total cost. The change from acid Bessemer to acid open-hearth steel raised total cost, and this process was used only in the special circumstances already described. The change from acid Bessemer to basic Bessemer should have lowered costs by this argument and should have been observed. This transition did not take place in the United States, however, the reason being that even though the two acid processes used the same kind of iron, the two basic processes did not. The above argument dealt with the raw materials for the two acid processes and for the basic open-hearth process; it did not discuss the inputs to the basic Bessemer process.

The acid process did not remove phosphorus from the iron, which meant that only a restricted class of iron ore, the non-phosphoric or Bessemer ores, could be used to make iron for conversion into steel. The use of a basic lining removed phosphorus, and the ores containing phosphorus could be used. Making a wider class of inputs suitable will not raise costs, and it may, as in the figures cited, lower them. But there is a complicating factor in the Bessemer process which meant that the new innovation altered, but did not increase, the range of suitable raw materials. The Bessemer process is conversion without fuel, that is, without an external supply of fuel; the combustible elements that provide the heat for the reactions come from within the iron itself. In the acid Bessemer process, this element is silicon which combines with the furnace lining in the course of burning. The lining is different in the basic process, and silicon will not combine with it. The lining is designed to attract phosphorus, and it is the heat generated by the combustion of phosphorus and its consequent combination with the furnace lining that generates the heat necessary for the basic Bessemer process. Phosphorus, in other words, went from being a liability to being a necessity. Iron containing more than 0.1 per cent phosphorus was not suitable for the acid Bessemer process; iron containing less than 1.5 per cent phosphorus was not suitable for the basic Bessemer process.[54]

The Bessemer process was permitted by the introduction of the basic process to use either of two very restricted classes of ores. The open-hearth process, which did not depend on the iron itself for the source of its heat, was enabled by the basic process to use ore with almost any proportion of phosphorus; the middle ranges of phosphorus were eliminated by the basic lining as easily as the larger proportions. The price of iron suitable for the open-hearth process dropped as a result of this innovation and produced the cost structure illustrated in Table 6.1. The price of iron suitable for the Bessemer process, however, did not drop and there was no corresponding advantage for the use of this process. There were very few ores in the United States suitable for the basic Bessemer process, and this process was never widely used.[55]

Bessemer or Open-Hearth Steel?

In the course of the 1880's, therefore, the costs of making steel by the basic open-hearth process approached those incurred by using the acid Bessemer process. This shift in the supply curve for open-hearth steel made the two types of steel competitive; two developments of the late nineteenth century then provided the small impetus necessary to tip the balance toward the open-hearth. On the supply side, the increasing availability of scrap lowered the costs of the open-hearth a little. On the demand side, the replacement of the demand for rails by the need for other products made of higher quality steel furthered the transition away from the converter. In most modern discussions, the three factors listed here are isolated as causing the change under discussion, but no attempt is made to allocate importance among them. We see clearly, however, that the ability of the basic open-hearth process to use the phosphoric ores of the United States was the most im-

[54] Clark, Vol. II, p. 267; Popplewell, pp. 16-17.
[55] Campbell, pp. 10-11; Popplewell, p. 32.

portant of these three factors. The other two—the differences between the two processes in their use of scrap and quality of product—were the marginal factors that determined the switch after the change in the supply curve for open-hearth steel had made it competitive.[56]

The Bessemer process was limited in the amount of scrap it could use for the same reason that it was limited in the range of suitable pig iron. The heat for the conversion was generated by the materials themselves, and they had to be of an appropriate composition to supply the necessary heat. The open-hearth process, on the other hand, with its freedom from this constraint, could use as much scrap as pig iron in its inputs, although too much scrap was harmful in this process also.[57] The importance of scrap in lowering the costs of the open-hearth process has prompted people to assert that the Bessemer process had to precede the open-hearth process to create the necessary scrap.[58] This is not evident in the price data, for although the price of scrap fell over this period, the price of pig iron did also, and there was little change in the relative price. "Busheling scrap" became the cheapest form of iron and steel scrap in the last years of the nineteenth century, but its price relative to that of Bessemer pig iron only declined from near 0.8 about 1890 to approximately 0.65 after the turn of the century. It was somewhat cheaper during some years of the depression, but other forms of scrap did not show a relative price decline.[59]

The price of scrap was thus falling rapidly enough to encourage its use, but not rapidly enough to be a major force in the restructuring of the steel industry. It would appear that the supply of scrap was expanding in these years at the same time as the demand for open-hearth steel was rising due to

[56] A representative listing of causes for the change to open-hearth steel may be found in Joseph Newton, *An Introduction to Metallurgy* (second edition; New York: John Wiley and Sons, Inc., 1947), p. 531. A better treatment, which uses essentially the same argument as the one above, can be found in Douglas Alan Fisher, *The Epic of Steel* (New York: Harper and Row, 1963), p. 129.

[57] *Iron Age* (December 12, 1878), p. 1; Iron and Steel Institute (Great Britain), *The Iron and Steel Institute in America* (London, 1891), pp. 271-281; U.S. Commissioner of Corporations, III, 153.

[58] H. W. Graham, *One Hundred Years* (Pittsburgh: Jones and Laughlin Steel Corporation, 1953), p. 14.

[59] *AISA* (1904), pp. 125-135. Other forms of scrap for which prices are given are "Heavy cast scrap" and "Railroad wrought scrap." Prices for scrap do not extend back beyond 1889.

the quality factors to be discussed shortly. The result was an approximately stable relative price for scrap, but with a tendency for it to decline.[60]

From the point of view of the individual steelmaker, who looked at the prices in the market as given, the importance of the changing supply of scrap was not very great. From the point of view of the historian, who is able to see the expansion of demand that exerted an upward push on the price of scrap, the expanding supply of scrap appears more significant. Nevertheless, scrap was not a necessary material in the production of open-hearth steel; other materials could have been substituted for it—as in the "pig and ore" process of Siemens—[61] if its price had risen. A static supply curve for scrap in the late nineteenth century would therefore probably not have prevented the extension of the open-hearth process.

Nevertheless, the close costs of the Bessemer and open-hearth process resulting from the use of the basic process and of scrap allowed small differences in the demand for the two types of steel to influence the composition of output. The change in demand for specific products will be examined in Chapter 10; we here discuss the difference in the quality of the various types of steel that allowed differences in the demand for products to be communicated into differences in the demand for types of steel. As before, the existence of acid and basic variants of both steelmaking processes forces us to discriminate between the effects of the process and of the lining.

Let us begin by quoting from a production manual written about the turn of the century:

> It is a common belief that it is an easy thing to distinguish between open-hearth steel and Bessemer steel. It is usually very easy to tell basic open-hearth steel from acid Bessemer, or acid open-hearth from basic Bessemer, but it is impossible by any ordinary means to tell acid Bessemer from acid open-hearth or basic Bessemer from basic open-hearth. Most American metallurgists and engineers, however, agree that open-hearth steel of a given composition is more

---

[60] These comments represent a rather casual solution to the identification problem implicit in the discussion. They are based on the assumption that the demand curve for scrap was a function of the production of open-hearth steel and that the supply curve was a function of the production of all steel. Both curves were probably fairly elastic over a wide range.

[61] Carr and Taplin, p. 34.

reliable, more uniform, and less liable to break in service than Bessemer steel of the same composition.[62]

This production manual also tells why this agreement exists: "Although there is no essential difference in the results obtained from open-hearth and Bessemer steel in the ordinary testing machines, there is good testimony to show that the product of the converter is an inferior metal which gives way in a treacherous manner under shock." To support his position, this author quotes the chief of a Pittsburgh testing laboratory: "Numerous cases have come under our observation of angles and plates which broke off short in punching, but although makers of Bessemer steel claim that this is just as likely to occur in open-hearth metal, we have as yet never seen an instance of this kind in open-hearth steel."[63] The conflict between Bessemer and open-hearth steel had by then moved beyond its original grounds of discussing the uniformity of carbon content, but had not yet come as far as identifying the cause of observed differences between the metals. The people who dealt with steel were in the uncomfortable position of observing differences without being able to explain them. The tendency of Bessemer steel to fracture unexpectedly made these differences of prime importance in some uses of steel and gave to their discussion its highly emotional cast. If one material was indistinguishable from another by "any ordinary means," but was liable to fracture without warning, how else would you describe it than "treacherous"?

But while most steel men agreed on the relative merits of open-hearth and Bessemer steel, there was no general agreement on acid and basic steel.[64] They were slightly different, and it is likely that each was better for some uses and that people generalized from the uses with which they were concerned. In addition, there was the problem of learning the characteristics of the newer basic steel, and this created some problems which may not have been closely related to the actual characteristics of the metal. Basic steel was greeted with skepticism when it was first introduced, but either its quality improved or people became less suspicious of it, and it won general acceptance.

[62] Campbell, pp. 14-15.

[63] Campbell, p. 529. The quote is from A. E. Hunt in the *Journal of the Iron and Steel Institute*, (1890), 316. Campbell was one of the ones trying to eliminate Bessemer steel from structural steel through classification. [*Iron Age, 55* (April 11, 1895), 768.]

[64] Campbell, pp. 24-25.

People continued to talk about it and to disagree about it, but without either the volume or the intensity of the discussion about Bessemer and open-hearth steel.[65]

The discussions of the carbon content, as well as of the phosphorus content, of steel of the late nineteenth century have to be seen in the light of the imperfect chemical knowledge of these years. The knowledge of materials and how they could be expected to behave was inexact, and the tools for measuring the chemical content of materials were imperfect. People tested their materials in somewhat impressionistic ways, both for physical and for chemical properties, and tried to build their findings into a unified description of the phenomena observed. The physical tests consisted of taking an ingot, hammering it or rolling it into the shape for which it was destined, and bending it or dropping a weight upon it to see what would happen. Dropping a five-ton weight on a rail from ten feet was a stiff test at the Troy steelworks in 1869.[66]

There were many different ways to test for chemical content, and they differed widely among themselves.[67] The result of matching the inexact chemical and physical data was an imperfect knowledge of the relationship between them.[68] In an effort to correct this situation, Holley was instrumental in the creation of a government-sponsored testing commission in the 1870's, but progress in this field of knowledge was slow.[69]

The debate on the quality of Bessemer and open-hearth steel continued throughout the last quarter of the nineteenth century. It was participated in by many steel men, and their views were highly correlated with the type of product their firm manufactured or worked with. Such comparisons of chemical content as were actually made do not look conclusive to the modern observer and do not appear to have influenced the discussion.[70] The motivation for the discussions was the fear

[65] See *Iron Age, 42* (December 6, 1888), 865; *Bulletin, 21* (October 26, 1887), 299; *29* (September 20, 1895), 210; *30* (March 20, 1896), 66.

[66] *Bulletin, 4* (October 13, 1869), 41; Abram S. Hewitt, "The Production of Iron and Steel in Its Economic and Social Relations," *Reports of the United States Commissioners to the Paris Exposition, 1867*, Vol. II (Washington, 1870), p. 33.

[67] Campbell, pp. 37-47; George E. Thackray, "A Comparison of Recent Phosphorus-Determinations in Steel," *AIME, 25* (October, 1895), 370-395).

[68] Campbell, p. 21.

[69] AIME, *Memorial . . .*, p. 57; *E & MJ, 19* (1875), 308.

[70] For example, H. M. Howe, "The Attainment of Uniformity in the Bessemer Process," *AIME, 15* (May, 1886), 340-354.

of fractures in Bessemer steel, and the disputants were more concerned with rationalizing or disproving their existence than with determining chemical content. We therefore find a supporter of the open-hearth process asserting that the more careful working of the steel in that process led to greater uniformity of product. Speed was the bane of the Bessemer works, enabling them to mix up a rail and structural steel lot, roll an ingot when it was too hot, or let a defective item slip through. Even if the chemical composition of the two steels was the same, this person contended, the more careful working of open-hearth steel gave it more desirable physical characteristics.[71] But we also find a champion of the Bessemer process, Captain William Jones of the Edgar Thomson Works, asserting that the large output of the Bessemer process required regularity of work and high quality machinery which led to a high-quality product. The very item which had been condemned for its evil effects by the open-hearth advocates was represented by the Bessemer men as the source of their strength.[72]

The conflict was obviously not to be settled by arguments such as these. It was not the average performance of the two steels that worried people, it was the chance of an extreme movement, a failure, on the part of Bessemer steel. As long as this suspicion was alive, no amount of discussion could argue it away. A wise comment on the contemporary discussion (by the author of the production manual quoted earlier) was that the average values for chemical content were not at issue, but that the extreme values were important. One bad example could ruin the reputation of a metal.[73] There does not appear to have been a difference in the variance of carbon content between the two steels, despite a modern tendency to ascribe the shift to such a difference.

This is where the discussion stood at the turn of the century. Steel men allowed their suspicions of Bessemer steel to sway their decisions, even though they were unable to rationalize them. Specifications were adopted for various classes of steel

[71] Alfred E. Hunt, "Some Recent Improvements in Open-Hearth Steel Practice," *AIME*, 16 (February, 1888), 693-728. This is the same Hunt quoted by Campbell on the merits of the two types of steel.

[72] *Iron Age* (May 12, 1881), p. 7.

[73] H. H. Campbell, "The Open-Hearth Process," *AIME*, 22 (August, 1893), 345-511.

at this time, and in sensitive areas, such as the formerly trouble-some axles and bridges, only open-hearth steel was acceptable.[74]

In modern discussion, attention has been completely trans-ferred from the carbon content, which apparently is as reliable in one kind of steel as another, to the properties of the steel depending on nonmetallic impurities. The inferiority of Bes-semer steel is attributed to two disabilities: lower quality control due to more rapid conversion, and difficulties stemming from greater nitrogen content.

Time is still needed to diagnose and treat the difficulties of specific lots of steel, and the open-hearth process supplies this time. The necessity for time seems odd in the light of modern science, but, according to a research supervisor of United States Steel, "With all the volume of discussion of this process [the open-hearth process] in the iron and steel literature, actual furnace operation by the smelter and others in the shop is con-ducted largely by empirical methods evolved from experience, with little benefit from physical chemistry or fuel engineering."[75] Important as this factor is in the present-day preference for open-hearth steel, however, the primitiveness of nineteenth-century quality controls makes it seem unlikely that the time of working was a crucial factor in the shift in steel production before the First World War.

Instead, the debilitating effects of nitrogen lay at the root of the discussions of that period. As one modern metallurgist has put it: "A considerable part of this criticism can be traced to the hardening and 'aging' of the steels associated with the higher nitrogen contents which are characteristic of Bessemer and duplex steels."[76] This factor was undiscovered at the turn of the century and gave rise to the perplexity recorded above. It was, however, a feature of Bessemer steel we may be assured was present, because it comes from the method of manufacture of the steel itself: "Since the major source of nitrogen in liquid steel is air, the control of nitrogen depends, to a large extent, upon the intimacy of contact of the liquid metal with air."[77] Open-hearth

[74] Campbell, *The Manufacture* . . ., pp. 550-582.

[75] B. M. Larsen, *A New Look at the Nature of the Open-Hearth Process* (New York: American Institute of Mining, Metallurgical and Petroleum Engineers, 1956), p. 3.

[76] D. P. Smith, L. W. Eastwood, D. J. Carney, and C. E. Sims, *Gases in Metals* (Cleveland: American Society for Metals, 1953), pp. 86-87. See also Graham, p. 34.

[77] Smith *et al.*, p. 86.

steel, therefore, benefited from its method of manufacture at least as much as from its duration. The use of an external supply of fuel in the manufacture of open-hearth steel resulted in less exposure to air, lower nitrogen content, and consequently better steel.

7

Economies of Scale and Integration

The Bessemer Process and Integration

The new processes described in the last chapter represented a drastic shift in the supply curve for steel in the years after the Civil War. The increased ease of producing steel resulted in a greatly expanded output of steel, an output that was composed in its early years largely of Bessemer steel destined for use in Bessemer steel rails.[1] The production of Bessemer steel had several characteristics that differed from those present in the production of wrought iron by puddling. This chapter and its successor are concerned with the implications of these characteristics. The present discussion deals with the operations of the steel-producing firms; Chapter 8 discusses the relations between firms.

The significance of the Bessemer process was not that it operated differently than other processes throughout its period of importance. By the end of the nineteenth century, all stages in the manufacture of steel products had become organized on a large scale, high heat was being used freely in many ways, and chemical sophistication was more or less present throughout the steel industry. The importance of the Bessemer process was that it contained these characteristics — large scale, high heat, rigid chemical requirements — from its inception onward. Blast furnaces, open-hearth furnaces, and rolling mills developed these attributes, but the Bessemer converter and its plant only increased them. A firm that adopted the Bessemer process at any point, therefore, had markedly different techni-

[1] Appendix C, Tables C.4-C.7.

cal properties than one producing wrought iron, while a firm that adopted the open-hearth process, or the most modern blast furnace or rolling mill practice in the 1870's was not very different in many regards from other firms. This chapter discusses the effects of the introduction of the Bessemer process and then shows how the properties of Bessemer plants were communicated to other parts of the producing operation.

It was observed in Chapter 6 that the Bessemer steel firms of 1880 were very much alike. One of their common attributes was their degree of integration: they all combined the operations of smelting and refining iron.[2] As the transformations of iron ore into pig iron and pig iron into wrought iron had been largely conducted by different firms before 1865, we may attribute the integration present in steel firms to the effects of the Bessemer process.

A start had been made along the path to integrated operations by the iron rail mills, the largest rolling mills of the ante bellum era. The primary stimulus for integration of these mills had been the change in the relative size of rolling mill operations resulting from the production of rails. A rail mill consumed the output of several blast furnaces, and the producers of rails built or bought blast furnaces in order to assure themselves of a satisfactory supply of pig iron.

The scale of an individual refining unit was increased enormously in the transition from puddling to the Bessemer process, and with it was increased the incentive for integration. Where a puddling furnace operated with units small enough for a man to lift, the Bessemer converter dealt with 5 tons or more of molten metal at a time. The speed of the latter process was greater than that of the former, and the volume of output per unit time was increased. The comparison of puddling with the open-hearth process is different; there the increase in the volume of metal treated as a unit was offset by the longer time needed to treat it, and there was no increase (initially) in the output per unit time in a transition from puddling to the open-hearth process. In 1880, the average capacity of an iron rolling mill was about 12,000 tons of iron a year. The capacity of a Bessemer steel works was about ten times as much, 114,000

[2] American Iron and Steel Association, *Directory to the Iron and Steel Works of the United States* (Philadelphia, 1876-1900), 1880.

tons, but the capacity of the average open-hearth steelworks was actually less, about 10,000 tons.[3]

A Bessemer steel plant, in other words, initially needed the output of many blast furnaces to supply its converters. The number of furnaces necessary declined over time as the output of blast furnaces advanced, and by the end of the century the ratio of steelworks to blast furnace was about the same as the Civil War ratio of rail mill to blast furnace: about one to three or four. While this was happening, however, Bessemer steel producers made strenuous efforts to control their sources of pig iron, and they felt they were better able to deal with the comparatively small blast-furnace units through ownership than through the intermediary of the market.

Control was more important for the user of the Bessemer converter than for the user of the puddling furnace because of the stringent chemical requirements of the Bessemer process. The rigid specifications on the composition of the pig iron used meant that Bessemer manufacturers had to obtain pig iron that varied less in quality than the pig iron available on the relatively ungraded market. This objective was achieved by direct management of the producing blast furnaces.

A separate advantage of unified ownership that was exploited by the makers of Bessemer steel was the opportunity to integrate operations. If all the operations in transforming iron ore into finished rolled steel products were performed in a single large plant, many economies were possible. They were of two types, deriving either from the use of surplus heat generated at one stage in this process to fuel another or from more direct linkages between the stages. Improvements in these areas were many and effective.

There was also a gain from integration deriving from the flow of information stimulated by combining all the operations under one management. It is not known in general whether information flows more easily across process lines than across firm boundaries, but the experience of the steel industry would indicate that this was so in the late nineteenth century. Two kinds of evidence may be cited. The first is that the integrated

[3] R. N. Grosse, "Determinants of the Size of Iron and Steel Firms in the United States, 1820-1880," unpublished doctoral dissertation, Harvard University, 1948, pp. 281-282.

steel firms were the most efficient producers at all stages of production. An integrated blast furnace was more likely to have participated in the cost-saving innovations of these years than a merchant (unintegrated) blast furnace. And the integrated furnace was more likely to have the services of trained scientific personnel than was its independent counterpart. The presence of a chemist was surely a benefit to any blast furnace, as were lower costs, and the localization of much of the technical revolution of the late nineteenth century in integrated firms can only be attributed to information flows.

The use of trained chemists in the manufacture of iron and steel is an innovation in this period that finds no analogue in the earlier transition to the new iron industry. The causes of this change were the increase in scientific knowledge, the rigid chemical requirements of the acid Bessemer process, and — although less important than the other factors — the problems raised by the impurity of bituminous coal relative to charcoal or anthracite.

The increase in scientific knowledge is outside the scope of this study. The increasing awareness of iron men in this area, however, is worth noting. A contemporary observer recorded that "until about 1875 the fact that iron smelting was a chemical process was not generally accepted."[4] It was appreciated, though, by a few leading iron and steel men. The Wyandotte company apparently employed a chemist briefly in the course of its experimentation with the "Kelly" process. He started an investigation of the effects of nitrogen on steel, but resigned before its completion.[5] After this abortive effort to use the services of a chemist, the first known attempt was by Carnegie at the Lucy furnace. A chemist was employed there before 1872, a move judged to be a success. This started the general use of chemists, their main job at a blast furnace being to keep the phosphorus content of the pig iron low enough to suit it for use in a Bessemer converter.[6]

[4] *Iron Age,* 57 (January 2, 1896), 21.

[5] W. F. Durfee, "The Manufacture of Steel," continued, *Popular Science Monthly,* 40 (November, 1891), 25.

[6] Andrew Carnegie, *Autobiography of Andrew Carnegie* (Boston: Houghton Mifflin, 1920), p. 175; James Howard Bridge, *The Inside History of the Carnegie Steel Company* (New York: Aldine Book Co., 1903), p. 65; Fritz Redlich, *History of American Business Leaders* (2 vols.; Ann Arbor, Michigan: Edwards Brothers, 1940), p. 81.

Chemists were probably used elsewhere at this time or before, by somewhat less loquacious people. The technical nature of iron and steel manufacture was recognized by the middle 1870's, even if the specifically chemical nature of it was not. The sponsorship of educational and research institutions by iron men was easily noted by Bell on his 1874 visit to the United States.[7]

The Integrated Blast Furnace

We now follow the operations of an integrated steel firm, from the production of pig iron to the rolling of salable products. The innovations discussed appear in roughly their chronological order, although the correspondence between order in time and in production is not complete. We are interested primarily in those innovations reflecting, or producing, gains from larger scale operations, increased attention to chemical matters, and greater integration of operations.

The key innovation in blast-furnace practice in these years was "hard driving," that is, the process of increasing the output of a given blast furnace over its rated capacity. The start of this process is variously dated at 1870, with the construction of the Lucy furnace, or at 1879, with the construction of the first Edgar Thomson Steel Works furnace. Whichever event is used, the important role of Andrew Carnegie in the development of the steel industry is illustrated, as he was an owner of both these furnaces. He was not a blast-furnace manager, of course, but the exploitation of the idea which was imported from England can be credited to him.[8] The introduction of accurate accounting techniques, which enabled the advantages of innovations to be seen, has also been claimed by him.[9]

David Thomas discovered that hotter blasts under higher pressure could produce more output economically before 1840. This discovery was of paramount importance in the introduction of mineral fuel into the blast furnace in America. For thirty

---

[7] I. Lowthian Bell, *Notes of a Visit to Coal and Iron Mines and Ironworks in the United States* (second edition; Newcastle-on-Tyne, 1875), pp. 32-33.

[8] See Victor S. Clark, *History of Manufactures in the United States* (3 vols.; New York: McGraw-Hill Book Company, Inc., 1929), Vol. II, p. 254.

[9] Carnegie, *Autobiography*, pp. 129-130; David Brody, *Steelworkers in America, the Nonunion Era* (Cambridge: Harvard University Press, 1960), pp. 18-19.

years or more after this initial discovery, however, no one thought of increasing the output of blast furnaces more by further increasing the heat and strength of the blast. Before 1879, according to one observer, "Increase of product was not thought of because it was not deemed wise or safe to attempt anything beyond the rated or recorded capacity, and the furnace 'seemed to be doing all it could, anyway.'"[10] This blindness was remarkable for two reasons. First, there was the similarity of hard driving to the innovation of Thomas; that the recorded capacity of blast furnaces should have become so quickly frozen at its new level is surprising. In addition, blast-furnace practice in England was expanding along just these lines. There, chiefly in the Cleveland region, blast furnaces were built between 1855 and 1875 with greater height, hotter blasts, and consequent larger yields. This idea penetrated to the United States during the 1870's at the same time as it was dying out in Britain. It would appear strange that at the same time the Americans began to import the techniques being used in Britain, the British ironmasters lost sight of the necessary balance between the various innovations introduced and abandoned their progress.[11]

One factor that may have induced the blindness in America was the difference between the fuel available in the United States and in Britain that was the concern of Chapter 3. Anthracite was well adapted for use in the ante bellum iron industry because of its chemical purity, but its physical structure resisted rapid combustion. It was harder to "drive" an anthracite furnace than a coke furnace, and the introduction of hard driving may have had to wait upon the exploitation of suitable sources of coke, a development of the Civil War and succeeding years. The direction of causation also worked in the opposite direction if it worked at all, and the introduction of hard driving encouraged the use of coke as a blast-furnace fuel. Almost all the blast furnaces to be considered in this chapter used coke for fuel; the merchant blast furnaces were the principal users of anthracite, and their problems will be considered separately.[12]

The first furnaces to be built in the Pittsburgh area conformed to a standard model that was not too different from the Lonacon-

---

[10] E. C. Potter, "Review of American Blast Furnace Practice," *AIME*, 23 (August, 1893), 370-382.
[11] J. C. Carr and W. Taplin, *History of the British Steel Industry* (Cambridge: Harvard University Press, 1962), pp. 50-53.
[12] See Chapter 9.

ing furnace. After the Civil War a furnace in Ohio began to attract attention to its large outputs, and the British innovations were recognized. In 1870, two blast-furnace companies were formed in Pittsburgh and a set of rival furnaces were built. These were the famous Lucy and Isabella furnaces, the former being the property of the interests associated with Carnegie. These furnaces were larger than previous furnaces, being 75 feet tall and 18 or 20 feet across the bosh. They began their contest of ever-increasing outputs upon their completion in 1872, the first example of a serious attempt in America to raise the output of a blast furnace on a sustained basis. The first Lucy furnace made 13,000 tons of iron in 1872, about twice as much as a very large furnace of a decade earlier. This amount was increased through various means to over 100,000 tons a year in the later 1890's.

The outputs of these furnaces interested the industry, but they were not immediately imitated. This delay is undoubtedly related to the initial delay in trying for higher yields, but is still not adequately explained. The next famous furnaces were the first two furnaces built at the Edgar Thomson Steel Works. The first was built in 1879 from the transported parts of a charcoal furnace originally located at Escanaba, Michigan. Under the management of Julian Kennedy, this furnace achieved far higher outputs at the Edgar Thomson works than it had at its original location. The next year a second furnace was built, the Edgar Thomson B furnace, and this furnace, built to the expanded specifications that were then becoming known, was sometimes regarded as the first of the new style furnaces.[13]

More furnaces were built with an eye toward increased outputs after these initial ventures had proved successful, and by 1890 there were several furnaces making over 1,000 tons of pig iron a week. This figure may be compared with the weekly output of 70 tons produced by the early coke furnaces and by the Lucy and Isabella furnaces at their inception. Or it may be compared with the yearly production achieved at one point by the Lucy furnace of over 100,000 tons a year, implying a weekly production of over 2,000 tons. The largest furnaces belonged to the two furnace companies already named and

[13] The foregoing discussion has been taken from Bridge, pp. 54-70, 87-90. Bridge appears to take much of his information from *Iron Age*, although he does not give explicit sources for most of his material. See *Iron Age* (January 31, 1884), p. 17; (February 28, 1884), p. 26; *46* (October 9, 1890), 571; *47* (April 23, 1891), 770.

to the following integrated Bessemer steelworks: Cambria, Edgar Thomson, South Chicago, and the Pennsylvania Steel Company.[14] It should not be supposed that these figures are typical of the industry. They represent the best practice, which was widespread only within one sector of the industry; average practice was far below them. The average daily capacity of blast furnaces in 1890 was given by the census as 68 tons, implying a weekly capacity of less than 500 tons.[15] This was far larger than the capacity of ante bellum furnaces, and it was increasing rapidly, but it was still less than half as large as the furnaces of the large steel companies. The scale-enlarging innovations of these firms filtered down only partially to the rest of the industry.

The innovations whose combination produced hard driving were three in number: larger blast-furnace sizes, more powerful blast engines, and hotter blasts. The increase in size required little more than a modification of ideas to be feasible. More powerful blast engines were produced by increasingly sophisticated technology, but they could have been built before, if not as cheaply, and their construction did not raise problems for the iron industry. The hotter blast, however, was a serious problem, because the original hot-blast stoves composed of cast iron could neither produce nor stand high heat. Innovation in this field was a necessary antecedent to hard driving.

Prior to the Civil War the blast had been heated by passing it through pipes placed in the path of the gases escaping from the top of the blast furnace. Shortly after the Civil War, the stoves invented by John Player of England were introduced into America. These stoves placed the pipes containing the blast on the ground, bringing the waste furnace gases to the blast rather than the blast to the waste gases. The pipes, however, were still made of cast iron, and the temperatures reached of about 1,000°F. represented the limits of tolerance of this material. The temperatures were nevertheless almost double those customary with the older hot-blast stoves.[16]

Shortly after the introduction of the Player stove, another

[14] Iron and Steel Institute (Great Britain), *The Iron and Steel Institute in America in 1890* (London, 1891), pp. 348-349.

[15] U.S. Census of Manufactures, 1905, p. 14, See Table 7.1.

[16] Clark, Vol. II, pp. 76-77; *Bulletin, 4* (October 13, 1896), 47; *Iron Age* (July 22, 1875), p. 1. The last-named reference contains a description of all the hot-blast stoves discussed here.

modification of the hot-blast stove was introduced. This was the Cowper stove, based on the regenerative principle of Siemens. This stove was made of firebrick and was therefore not as subject to heat-induced deterioration as the Player stove. The temperatures attained, however, were not markedly greater than those attained with the Player stoves, and the regenerative stoves did not completely supersede the cast iron ones. Various modifications of Cowper's initial stove were introduced, labeled by the appropriate series of hyphenated names, and a new concept was introduced by Whitwell. He used a series of interlocking walls to contain the firebrick, instead of the checkerwork used by Cowper. There would seem to be obvious economies in cleaning costs with this more accessible form, but there must have been offsetting disadvantages (possibly in the heat retention of the stoves), because it was not widely adopted. Regenerative stoves in any form, in fact, were not widely adopted before 1885, although they spread extensively thereafter.[17]

The temperature of the blast was therefore doubled in the decades following the Civil War, and the pressure of the blast underwent a similar increase in intensity. The highest pressure before the war was something over two pounds per square inch. A critical innovation of the Lucy and Isabella furnaces was their use of a very high pressure: eight or nine pounds per square inch. This pressure represented the beginning of a divergence between British and American blast-furnace practice, the former keeping a blast of four to five pounds pressure. By 1890 this high pressure was typical of advanced American blast furnaces, and by the turn of the century the blast pressure in use had risen even higher. Part of this gain was due to the use of more powerful blowing engines, but most of it was due to the use of more blowing engines, one for each furnace in place of one for three or four.[18]

It will be noticed that many of the references for the preceding statements are British sources and that many of the innovations discussed were originally British innovations. The

[17] Iron and Steel Institute, pp. 348-349; Carr and Taplin, pp. 50-51; *Bulletin*, 5 (October 26, 1870), 57; *12* (July 10 and 17, 1878), 161; *12* (August 7, 1878), 180; *Iron Age*, 57 (January 2, 1896), 21-24; J. Stephen Jeans (editor), *American Industrial Conditions and Competition* (London, 1902), p. 442.

[18] Bell, pp. 37-38; Iron and Steel Institute, pp. 348-349; Jeans, p. 483; Carr and Taplin, p. 53.

nationality of the innovations was the cause of some confusion on the part of British ironmasters as they watched the best efforts of America surpass theirs, and the discussions just cited were a result of this concern. There were many attempts in the last quarter of the century to rationalize the British practice, explain away the American practice as irrational, or, occasionally, to discover why the two countries were different. The discussions were posed in terms of the most advanced technology of the countries, rather than the average level of achievement, partly as a matter of local pride, but partly also because the best practice was typical for large integrated steel firms, even if not for the iron industry as a whole.

Duncan Burn summarizes the results of the discussion and of the international differences in the rate of technological change in his history of the British steel industry:

> The United States became the acknowledged pioneer in blast-furnace practice, developing along the lines laid down in the seventies but harmonising them with economy both of fuel (a Chicago feat) and of furnace linings, and crowning the work with mechanical charging whereby coke and ore for the furnace were filled into buckets or skips from bunkers at ground-level, weighed and carried to the top of the furnace, and then emptied from the buckets and distributed suitably in the stack, wholly by mechanical means controlled by an operator working at the foot of the furnace.[19]

Burn notes that this leadership of the American industry was tightly linked to the large outputs attained by American blast furnaces. Large outputs permitted the use of large machinery, such as the skip-hoists noted in the quote and casting machines for the pigs, and the use of these machines required high outputs as the capital charges would overwhelm any furnace that did not make full use of them. The British industry was not characterized by large outputs per firm and was not able to take advantage of the American extrapolations from the British innovations of the 1870's and before.[20]

British ironmasters argued, in defense of their actions, that hard driving was not profitable. Whether or not this was true

---

[19] D. L. Burn, *The Economic History of Steelmaking, 1867-1939* (Cambridge, England: At the University Press, 1940), pp. 183-184. Burn discusses the international comparison at some length in his Chapter 10, pp. 183-218, where he documents the contemporary discussion of the problem.

[20] *Ibid.*, pp. 188-192.

in England, it was not true in America. We asserted in the previous chapter that there were economies of scale in the manufacture of Bessemer steel sufficient to motivate the expansion of plant size in the area. The evidence presented above provides similar indirect support for the existence of economies of scale in the manufacture of pig iron. The scale of production of a given blast furnace could be greatly increased by the use of several relatively inexpensive devices for increasing the heat and the pressure of the blast. Since labor was specific to the furnace rather than to the volume of production, the labor cost declined with the capital costs. For small changes the improvements were undoubtedly profitable because of the already large volume of investment. The investments represented by extensive auxiliary equipment to handle material and the patented machines for preparation were larger, and a cost calculation would be needed to demonstrate their advantages.[21] Without this proof, it appears that there were economies to be gained by blast furnaces through these innovations.

In the course of the last two decades of the century, also, the openings of the blast furnace were closed by the Lurmann front and the bell and hopper (later the double bell and hopper) top, pneumatic lifts and skip-hoists reduced or eliminated the labor of charging the furnace, and casting machines eliminated that labor in taking the product from the furnace that was not eliminated by the direct process. The shape of the furnace, the "lines," was altered to achieve greater yields and fuel economy, but the contemporary discussions give evidence of continuing ignorance of the optimum shape.[22]

Steelworks and Rolling Mills

The first of the economies resulting from integrated operations was the use of the "direct process." The problem with the direct

[21] But see Table 7.1.

[22] For accounts of the technical aspects of these innovations, see James Gayley, "The Development of American Blast-Furnaces, with Special Reference to Large Yields," *AIME, 19* (October, 1890), 932-995; Axel Sahlin, "The Handling of Materials at the Blast-Furnace," *AIME, 27* (February, 1896), 3-42; Louis Grammer, "A Decade of American Blast-Furnace Practice," *AIME, 35* (February, 1904), 124-139; Richard Peters, Jr., *Two Centuries of Iron Smelting in Pennsylvania* (Philadelphia: Pulaski Iron Company, 1921). The last-named source has fine pictures of blast furnaces as they were at many points during the nineteenth century.

process, that is, with running molten pig iron directly from the blast furnace to the Bessemer converter, was that the output of a blast furnace varied too much for the highly sensitive converter. The process was used at a few mills, but not widely.[23] By 1890, Captain Jones had expanded the use of the intermediate ladle into a mixer by greatly enlarging it and making it into a reservoir of molten pig iron. The converter charge taken out of it had an even quality because the mixer was large enough to average out the fluctuations in quality from blast-furnace operations. The Jones mixer was used at the Edgar Thomson works in 1890, and it had spread throughout the industry by 1900.[24]

From the mixer, or from remelting in a reverberatory or cupola furnace, the metal was supplied to the Bessemer converter. The speed at which steel was made was continually rising, and new innovations were constantly being introduced to speed it further. We may let Clark survey the changing aspect of Bessemer plants.

> For fifteen years the typical American Bessemer plant was designed along essentially the lines adopted by A. L. Holley when he installed the first permanent apparatus at Troy; but between 1879 and 1882, a notable change took place in plant arrangement, with the object of enlarging output and saving labor. The so-called American lay-out had hitherto consisted of two converters of about five-tons capacity, side by side, served by a single ladle crane . . . The changes which now occurred were in two directions: converters were enlarged and their number was increased to permit the use of two or more ladle cranes, so that the three and four-converter plant with seven-ton to ten-ton converters speedily became the standard arrangement in America.[25]

The limit on speed in the converting process soon became the speed at which ingots could be removed from in front of the converters. This was solved partially by the use of large

---

[23] Abram S. Hewitt, "The Production of Iron and Steel in Its Economic and Social Relations," *Reports of the United States Commissioners to the Paris Exposition, 1867* (Washington, 1870), Vol. II, p. 66; *Bulletin, 16* (February 22, 1882), 57; *16* (November 8, 1882), 298; *17* (January 17, 1883), 12.

[24] *Iron Age, 46* (September 18, 1890), 440-443; Harry Huse Campbell, *The Manufacture and Properties of Iron and Steel* (second edition; New York: The Engineering and Mining Journal, 1903), 169-170; Frank Popplewell, *Some Modern Conditions and Recent Developments in Iron and Steel Production in America* (Manchester: At the University Press, 1906), pp. 93-94; Jeans, p. 511.

[25] Clark, Vol. II, pp. 265-266.

converters, and therefore of fewer ingots, and partially by a characteristically American device of casting the ingots directly on trucks and hauling them away by locomotives without further ado.[26]

The practice was initially to let the ingot cool completely and then to heat it for rolling, as would be done in an unintegrated process. The combination of converting and rolling presented the possibility of using the initial heat of the ingot for rolling. This was first done in England around 1880, and the practice spread rapidly to the United States. The enabling device was the "soaking pit" in which the hot ingot was placed and allowed to cool slowly, and therefore uniformly, to a rolling heat.[27]

Once the ingot reached the rolling mill, it was treated by a multitude of machines, which the large and continuous volume of inputs enabled the rolling mill to utilize efficiently. It was moved around on the carrying rollers introduced by the Fritz brothers—from the blooming tables invented by George to the three-high mills designed by John—this movement being powered by engines also designed or adapted by these extraordinary brothers.[28] As a result of these and other innovations similar in spirit by Jones, Hunt, and others, steam and later electric power replaced the lifting and carrying action of human muscle, mills were modified to handle the steel quickly and with a minimum of strain to the machinery, and people disappeared from the mills. By the turn of the century, there were not a dozen men on the floor of a mill rolling 3,000 tons a day, or as much as a Pittsburgh rolling mill of 1850 rolled in a year.[29] By arguments similar to those used for blast furnaces, we may conclude, barring actual cost calculations, that the efficiency of rolling mills was increased by the use of these devices and that their greater suitability for integrated than for independent operations implied a greater efficiency for the integrated mills.

[26] Bridge, p. 109.

[27] *Iron Age* (October 26, 1882), p. 13; *Bulletin, 17* (January 17, 1883), 12; *17* (August 8, 1883), 213.

[28] *E&MJ, 16* (1873), 153.

[29] Popplewell, p. 103. Popplewell notes (p. 71) that Carnegie would spend $1,000 on a machine to replace a man. If this is a maximum figure, it implies a very high rate of carrying charges for the machine, due to either rapid obsolescence or a high level of breakage. Clark, Vol. II, pp. 252-254, notes the temporary use of natural gas in rolling mills and steelworks in the late 1880's.

Although cost data are not readily available, a certain amount of data bearing on technical efficiency has been preserved in the censuses of the period. We may summarize some of these data in a form similar to that used in Table 5.1 for the ante bellum iron industry. The data in Table 7.1 are for the entire country rather than for Pennsylvania, and for a series of years rather than one. But they represent, perhaps, the same order of approximation as the older figures do.

The entries for 1869 are of approximately the same magnitude as the data given in Chapters 4 and 5 for the largest establishments before 1865. The largest ante bellum plant had become the average size plant of 1869, and plant size continued to grow, the largest plants being always well above the average. These data do not reflect the degree of integration among steel firms as the census collected separate data for blast furnaces independent of their ownership. The size of a large integrated firm was much larger than the size shown here for individual establishments.

TABLE 7.1
AVERAGE SIZE OF IRON AND STEEL ESTABLISHMENTS, 1869-1899

	Capital (thousands of $)	Census-Year Product (thousands of tons)	Wage Earners (no. of persons)
Blast furnaces			
1869	145	5	71
1879	262	10	122
1889	426	29	110
1899	643	65	176
Steelworks and rolling mills			
1869	156	3	119
1879	267	7	220
1889	661	14	332
1899	967	23	412

Source: U.S. Census of Manufactures, 1914, *Abstract,* pp. 640-641; Appendix C, Table C.8.

The average capital invested in both branches of the industry rose considerably faster than the average employment. The increase in output per employee which was the subject of so much comment at the time was therefore at least partly a substitution of capital for labor. Output per dollar of capital did not decline, however, and there was a gain in productivity. Output per dollar of capital rose for blast furnaces, indicating that the innovations chronicled above represented a reduction in the capital requirement per ton of pig iron produced as well as a reduction in the labor requirement. This was not true for steelworks and rolling mills, the changes there saving labor without reducing capital costs. These statements relate to productivity changes; to move from them to profitability statements would require wage rates, material costs, and capital charges.[30] Nevertheless, there would seem to be a presumption that these changes, since they reduced inputs, reduced costs. And it would also seem clear that there was at least as much inducement to increase the scale of blast-furnace operations to make use of these innovations as there was to enlarge steelworks.

As a result of these inducements, the growth of the iron and steel industry in the late nineteenth century was accomplished through a growth in the size of plants rather than through an expansion in the number of plants. In the years covered by Table 7.1, the number of blast-furnace establishments fell from 386 to 223, and the number of establishments classified as steelworks or rolling mills stayed near 440 without showing any trend.[31] The number of steelworks and rolling mills was the same as the number of establishments, but the number of blast furnaces was higher due to the existence of clusters of furnaces. The number of blast furnaces did not show the even trend reported for the number of blast-furnace establishments; it exhibited a pattern that illustrates a parallel between the transition from the old iron industry to the new and from the new iron industry to the steel industry. The number of blast furnaces had declined to a low of under 300 in the first transition, the increasing production of iron being more than offset by the increasing scale of production due to the adoption of anthracite. The number of furnaces then rose from 1860 to 1880, reaching a maximum of close to 700. From this level the number of furnaces declined to 400 at the end of the century

[30] Raw material consumption declined in this period. See Walter Isard, "Some Locational Factors in the Iron and Steel Industry since the Early Nineteenth Century," *Journal of Political Economy*, 56 (June, 1948), 203-217.

[31] See the sources for Table 7.1.

as ironmasters utilized the possibilities of coke to increase the size of the furnace faster than the volume of total production was growing.[32]

The largest iron rail mills had been supplied by three or four blast furnaces at the time of the Civil War. The steelworks initially represented a large jump in the intake of iron and increased the number of blast furnaces necessary to supply a rolling mill and associated activities. The extra supplies were bought on the market, but the 1870's and 1880's witnessed an expansion of pig iron producing capacity on the part of the steel firms as they tried to produce all their own pig iron. By 1890, the ratio of blast-furnace output to the output of the largest rolling mills had reached the level it had been before the age of steel. The largest Bessemer steel companies of that time owned up to fourteen blast furnaces, but up to four steel mills. They were the largest producers of pig iron in the country, and they made close to half the pig iron made in the United States.[33]

This sequence is reflected in the faster growth of blast-furnace size than of rolling mill size shown in Table 7.1. The table also illustrates the fact that almost all the innovations in the iron and steel industry in the nineteenth century increased the optimum plant size, the relation of different plants being determined by which grew the fastest. The only important exception to this trend was the open-hearth process, which could be conducted efficiently on a much smaller scale than the Bessemer process. The arguments used here to show why the steel mills owned blast furnaces did not apply to open-hearth plants, and it might be thought that the change from Bessemer to open-hearth steel would produce a decline in integration. Table 7.1 shows that by the time this transition took place, the size of blast furnaces had increased enough to make applicable an argument for integration that starts from the large size of these furnaces. And, in addition, the process of integration had reached an irreversible point by the 1890's, and the trend was toward more combination than can be explained on the grounds used thus far. This movement is one of the principal concerns of the next chapter.

---

[32] See the sources for Appendix C, Table C.8.
[33] Clark, Vol. II, pp. 234-235, 284-285; *Iron Age, 41* (January 19, 1888), 104.

8

Economies of Scale
and Industrial Organization

The Origins of The Bessemer Steel Industry

Over the late nineteenth century as a whole, the size of the producing unit in the iron and steel industry did not increase much faster than the size of the market. The discussion of the last section showed that the change in the number of establishments was not excessive. This over-all view, however, masks the importance of short period movements, the most important of which was the sudden and spectacular increase of plant size attendant upon the introduction of the Bessemer process. This process was already much larger than its competitors at the time of its introduction; one Bessemer plant could replace something on the order of ten iron rolling mills. The number of Bessemer plants built at the inception of the industry was therefore very small, and the conditions for free competition did not exist.

Despite the fact that the large minimum economic plant size limited the number of entrants to the Bessemer steel industry, Bessemer manufacturers tried after a while, to restrict entry still further by means of patent restrictions and other devices. This movement did not begin until almost a dozen firms had entered the industry, and when the era of unrestricted entry had ended, these firms were not able to agree on collusive measures for more than a few years at a time. Combinations were plentiful in the steel industry; the questions to be answered concern their effectiveness.

There were no productive processes in the ante bellum era that contained economies of scale similar to those of the Bessemer process or hard-driven blast furnaces, and there were

169

no industrial combinations on the scale of those found in the later steel industry. But if there was no effective combination, there was also no need to defend against accusations of combinations, and most of the surviving sources are reliable to the extent of their knowledge. This is not true of the post-Civil War era when the steel manufacturers were under continual fire for their restrictive policies. They evaded the issue by the creation of what has been called a "folk tale," describing the history of the industry in unobjectionable terms. This folk tale may have been the result of a conscious effort to confuse the issue, but it also derived in part from the unconscious efforts of reminiscing authors to recreate the past in a better form. The story, consisting largely of altruistic actions on the part of generous businessmen interested solely in the glory and growth of their country, cannot be relied upon. It is necessary to recreate the description of many choices made in the industry if a narrative is to make economic sense.[1]

We may begin the narrative in 1866, at the point when the two competing patent-owning groups had united and decided on a procedure for attracting licensees. They already had one licensee—the Pennsylvania Steel Company, a subsidiary of the Pennsylvania Railroad. For most of the following decade, the trustees of the patents and, later, the Pneumatic Steel Association attempted to enlarge the ranks of the Bessemer steel producers. In the remaining years of the 1860's, however, only one other Bessemer steel mill was opened. The Cleveland Rolling Mill Company, under the presidency of A. B. Stone with his interests in both railroads and rail mills, began to make Bessemer steel in 1868.[2]

This slow rate of entry was a result of low profit opportunities in the production of Bessemer steel in America with the

---

[1] The term "folk tale" was used by Elting E. Morison in conversation with the author. The folk tale itself pervades the writings of the nineteenth-century steel men and is most clearly articulated in those of James M. Swank. A good, brief example may be found in *Bulletin*, 25 (January 21 and 28, 1891), 20. W. Paul Strassmann, *Risk and Technological Innovation* (Ithaca: Cornell University Press, 1959), pp. 35-36, takes Swank to task for his distortion of the facts.

[2] This firm owned two iron rail mills before the Bessemer process was made available; Samuel Harries Daddow and Benjamin Bannan, *Coal, Iron, and Oil, or the Practical American Miner* (Pottsville, Pennsylvania, 1866), pp. 694-695. Stone was a director of the Lake Shore and Michigan Southern Railroad also; *Dictionary of American Biography* (New York: Charles Scribner's Sons, 1928-1936), Vol. XVIII, pp. 70-71.

technology then in use. The large size of a Bessemer plant meant that it was necessary to enter the industry whole-heartedly —the experience of the Freedom Iron Works standing as a reminder to the unbeliever—and people were willing to venture the required amount of funds only on the hope of considerable profit. This hope did not exist until Holley solved the problems of containing hot metal by the introduction of the Holley bottom at the end of the 1860's. At this point, with most technical problems solved or at least understood, the expectations of profits increased, and more firms entered the industry.

Five new mills started production in the years 1871-1873, and three more entered in 1875-1876. Of the five mills in the first group, three were built by people who were already connected with the industry. The Cleveland Rolling Mill Company built another plant in Chicago, the Union Steel Company, and two members of the old Kelly Process Company built plants. E. B. Ward and his associates built the Bessemer plant of the North Chicago Rolling Mill Company, and Morrell, one of the trustees of the joint patents, built a steelworks at Cambria. The two new entrants to the industry were the Joliet Iron and Steel Company, a new firm formed to buy and use the machinery of the Freedom works, and the Bethlehem Iron Company, the firm which employed John Fritz. Of the three new plants built in 1875-1876, all were new to the industry. They were the Edgar Thomson, Lackawanna, and Vulcan works, known respectively for containing Andrew Carnegie, containing W. W. Scranton, and being in St. Louis.[3]

The last three steelworks started production after the depression of the 1870's had begun, although all but the last had been organized earlier. Judging from the comments the steel men made a little later to defend their actions, they began to worry about overproduction soon after 1873. The small number of firms in the industry had been decried when times were good, but it looked like an unexpectedly good thing when times were bad. We would like to know if a change in profitability caused this change in attitude, or whether it was due entirely to a change in expectations. As we do not know the level of profits when new entrants could not be attracted,

[3] A list of these steelworks is in James M. Swank, *History of the Manufacture of Iron in All Ages* (second edition; Philadelphia, 1892), pp. 411-412. Details of their ownership are taken from American Iron and Steel Association, *Directory to the Iron and Steel Works of the United States* (Philadelphia, 1876-1900), 1880.

we cannot say much about its relation to the level when new entrants were not desired. But we can say a little about the level of profits during the depression of the 1870's.

It would appear from indirect evidence, largely comments by many people, that Carnegie's companies were among the most successful of the steel firms. Holley liked the Edgar Thomson Steel Works, built by Carnegie and his associates in the early 1870's, best of all the works he helped to build. He had started from scratch and let form follow function, and by that time the functions of a steel plant were quite well known.[4] This plant, as improved upon by the innovations of Captain Jones and others, was consistently a low-cost producer, and we may assume that its profits were the highest in the industry. Bridge reports that the profits were $42,000 for the three months in 1875 that the plant was operating, $181,000 for 1876, $190,000 for 1877, and $402,000 for 1878. The paid-in capital grew from about $750,000 at the start of production in 1875 to $1,250,000 in 1878. The precise times at which more capital was paid in is not clear; Carnegie subscribed much of the new capital and paid for it out of his earnings. If we assume that capital was $750,000 in 1876, $1,000,000 in 1877, and $1,250,000 in 1878, the profit rates for these years were approximately 25 per cent, 20 per cent, and 30 per cent.[5]

These profits were not earned by every firm in the industry, and some firms were not making any profits at all. The Joliet Steel Company failed twice in these years, once in 1874 and again in 1877. However, A. B. Meeker was in charge of the company both before and after the failures, and the legal procedures may have been only events in the course of a power struggle between Meeker and other groups within the company.

[4] American Institute of Mining Engineers, *Memorial of Alexander Lyman Holley* (New York, 1884), p. 135.

[5] James Howard Bridge, *The Inside Story of the Carnegie Steel Company* (New York: Aldine Book Co., 1903), pp. 94-101. Bridge states that his sources are the financial reports of the Edgar Thomson Steel Company, Limited, the operating company of the steelworks, but I have been unable to locate these reports. More information on the capital invested may be found in *The Papers of Andrew Carnegie*, Manuscript Division, Library of Congress, Vol. IV, and Burton J. Hendrick, *The Life of Andrew Carnegie* (Garden City, New York: Doubleday, Doran and Co., 1932), Vol. I. pp. 214-215. Herbert N. Casson, *The Romance of Steel* (New York: A. S. Barnes and Co., 1907), p. 26, repeats Bridge's profit figures, but makes a mistake for 1878 and 1879.

We simply do not know.[6] The Vulcan Iron Works, the last Bessemer steelworks to be built before 1880, was also in financial difficulty in these years, and we will return to its story in a moment.

If we conclude from this sparse evidence that the average profits in the industry were in the range of 10 to 20 per cent of invested capital, this means that the profits of the industry were typical of a new industry in the process of getting established. They were undoubtedly higher than the average industrial profits during the depressed years of the late 1870's, a result of the special features helping the steel industry's profit level, that is, economies of scale and the high tariff, and low profit rates do not appear to have been the stimulus for the steel industry to add other limitations of entry.

The economies of scale in steel production have been described earlier, and they need no further elaboration here. The tariff was an additional factor tending to increase the profitability of steelmaking. Before 1870, steel rails were classified as "manufactures of steel not otherwise provided for" and subjected to a duty of 45 per cent. This was changed to a specific duty of $28 a gross ton in 1870, a duty of approximately the same magnitude as the earlier *ad valorem* one. After 1873, however, the price of Bessemer steel fell sharply, and by 1877 the duty was the equivalent of about 100 per cent.[7] This duty had an increasingly restrictive effect as the decade wore on, and should have had a favorable effect on the profits of American manufacturers. The price of steel was declining sharply in these years due to technological change and to conditions in the depression. Profits would change as a result of the relative movements of the resultant supply and demand curves, and although we may say that the increasingly heavy tariff increased the demand relative to the available supply, we may not conclude from this that it was enough to raise profits.

The peculiarity of the tariff is that it was supported, not opposed, by the purchasers of steel rails. The railroads were

[6] See AISA, *The Duty on Steel Rails* (Philadelphia, 1880), p. 38; *Bulletin*, 7 (March 12, 1873), 216; 8 (July 9, 1874), 212; 8 (September 10, 1874), 267; 11 (April 4, 1877), 92; *Iron Age* (March 29, 1877), p. 11; (June 23, 1881), p. 9.
[7] F. W. Taussig, *The Tariff History of the United States* (fourth edition; New York, 1898), pp. 221-222.

actually in favor of increasing the duty on rails, and they demonstrated this by a letter to Congress asking for a higher duty.[8]
This letter may be interpreted, following Swank, as a reflection
of the railroads' fear of domination by English monopolies,
but it seems more accurate to regard it as the reflection of a
community of interests between railroad men and steel men.
In many cases, as for instance when the railroad man was
Samuel M. Felton who also happened to be the President of
the Pennsylvania Steel Company, this community resolved
into identity. As long as the costs to the railroad of new rails
could be passed on to the consumer, there was no reason to
balk at higher prices if they facilitated the growth of an industry in which the railroads were also involved.

Eight of the eleven Bessemer steel mills built before 1880
were added to existing rail mills,[9] while the remaining three
had special ties to the railroads. In fact, all of the Bessemer
plants had ties of one sort or another with the railroads, usually
through the medium of common ownership or directorships.
The Bethlehem Iron Company shared directors with the Lehigh
Valley and other railroads; the Lake Shore and Michigan
Southern Railroad was connected to the Cleveland Rolling
Mill Company and the Union Steel Company through their
president, A. B. Stone. Of the three new firms, the Joliet Steel
Company was connected to several Chicago railroads through
the connections of F. E. Hinckley and A. B. Meeker; the Pennsylvania Steel Company was one-third owned by the Pennsylvania Railroad; and the Edgar Thomson Steel Works was
connected to this same railroad through the good offices of
Andrew Carnegie, a former employee of the railroad, John
Scott, a director of the railroad, and Edgar Thomson, president
of the railroad until his death in 1874 and friendly to the firm
named after him even if he had declined to join it.

The Second Decade

The first attempt to organize the steel industry and restrict
entry took place in 1875, when the Bessemer Steel Association,
which continued in various forms for the rest of the century,

[8] *Bulletin,* 4 (January 19, 1870), 153; U.S., Congress, Senate Report No. 2332.
50th Cong., 1st Sess., 1888, III, 645-647.
[9] Swank, pp. 411-412.

was created. The meetings of 1875 attempted to form a pool, that is, an agreement to share the market for steel rails according to prearranged quotas. The story of Carnegie's indignant reaction when he discovered he had been allocated a smaller share than the other steel firms is well known.[10] The story asserts that Carnegie was given a large share and that the pool went into effect. But it appears in a letter from Carnegie to W. P. Shinn, the secretary and treasurer of the Edgar Thomson Steel Company, that lasting combination was not achieved. Writing on November 20, 1875, Carnegie said, "Of course we are in for a year of low prices. I have seen this from the day of our failure to combine, but if we meet it rightly the track will be clearer after the war is over—One year without dividend on Cambria, Penna. Steel, Joliet and North Chicago will make some amicable arrangement possible—bear in mind that with Cambria secured we could today have fine prices in the West. Our efforts should be mainly directed to this end."[11]

A more successful attempt at combination took place in 1877. In October, 1877, the steel men renewed their pool with penalties for violations, decided to restrict the availability of their patents to people currently manufacturing Bessemer steel, and vested these patents in the Bessemer Steel Company, Limited, a company formed for this express purpose and owned jointly by the eleven existing steel firms. When the *New York Tribune* accused the industry of forming a monopoly, Morrell pleaded fears of overproduction and Joseph Wharton of Bethlehem said the steelmakers had been "fleeced by the public" and that no agreement had been reached at the meeting, anyway.[12]

Nevertheless, the advertisements for licenses to the Bessemer patents, which had been in the *Bulletin* every week since February 27, 1867, appeared for the last time on October 3, 1877.[13] The editors of *Iron Age* commented on the situation in the steel industry just after the advertisements for licensees disappeared: "There are but eleven Bessemer mills in this country. They own absolutely all the patents essential to the

[10] See Hendrick, Vol. I, pp. 211-212.
[11] Carnegie MSS, IV.
[12] *E&MJ*, 24 (1877), 301; *Bulletin*, 11 (November 7, 1877), 289; 11 (November 21 and 28, 1877), 309.
[13] *Bulletin*, 11 (October 3, 1877), 263.

manufacture of their products. Put the price where they choose, the only competition will be among themselves, for no other mill can be started in opposition to them."[14]

But what patents were still valid? The original Bessemer and Mushet patents had worn out in 1870, and Kelly's patent rights had only been extended to 1878.[15] The patents referred to in *Iron Age* were newer patents, and the most important among them were undoubtedly Holley's patents discussed in Chapter 6.

The effectiveness of the new agreement is problematical, particularly since *Iron Age* felt impelled to comment shortly after their original article that they did not understand the low price of Bessemer steel in view of the existence of a monopoly by property right of the existing mills.[16] There was also not a fixed price for steel, and the price of steel products, which may be taken to mean rails, fluctuated between $41 and $43 from October, 1877, until the demand for rails started to rise sharply in July, 1879.[17] The conclusion to be drawn from this is that there was not a determined effort to fix the price of steel, either because the agreement of 1877 was not strong enough to permit it or because the steel men thought a limitation on output would be sufficient to keep up profits. The agreement was effective, then, to the extent that it limited the output of steel. To see its effectiveness in this area, we follow the development of the industry, with special attention to their actions toward the Vulcan and Homestead steel companies, the former being the weakest member of the rail pool and the latter an unsuccessful entrant to the industry in 1881.

According to Samuel Felton, the Bessemer Steel Company was formed solely for the purpose of assigning royalties for the patents it held. The capital of the company was $825,000

[14] *Iron Age* (November 8, 1877), p. 15. This was before the formation of the Bessemer Steel Company, implying that the new firm was a formalization of an agreement reached slightly earlier to restrict the availability of the existing patents.

[15] U.S. Patent Office, *Decisions of the Commissioner of Patents, 1870*, pp. 9-10, 106-108; *1871*, pp. 186-187; (Washington, 1871-1872). On Kelley's application, the Commissioner said that Kelly had "an expenditure of $11,500 in behalf of the patent, and receipts only to the amount of $2,400, leaving the patent indebted to him, at the date of its expiration, to the amount of $9,100" (*1871*, pp. 186-187).

[16] *Iron Age* (January 24, 1878), p. 14.

[17] *AISA* (1904), p. 121. See p.186 for more discussion of price stability.

and equal to the cost of the patents to the members of the company. The company did not fix prices, nor did it exclude new mills, which, however, would have to pay higher royalties to existing producers. It was owned equally by the eleven steel firms, and its profits were therefore divided equally among the member firms. Since royalties were collected according to the tonnage produced, this procedure amounted to a subsidy of the small companies by the large. The "generosity" of the larger members of the industry in fact exceeded this, according to Felton, for they paid the interest on the debts of the Vulcan Iron Works to keep them from bankruptcy during the depression. The Vulcan works were closed for "something over a year" during which the other mills paid $70,000 interest on a mortgage of about a million dollars. When demand was high enough for Vulcan to start production again, the other mills notified the closed firm and stopped the subsidy. Felton and Shinn, then managing Vulcan, both denied that the other steel mills controlled the Vulcan Works or ordered them to do things.[18]

This account is not altogether convincing, there being little motive for this generosity of the steel firms toward their unlucky cousin. Congressman Morrison asserted that the Bessemer Steel Company had "hired one of its associate establishments to put out its fires and stand still to maintain prices, while its laboring people were turned out of employment, to subsist on savings never saved."[19] The truth lies, no doubt, between these two extremes and may have been something like this. The Vulcan Works were the last to be built, being the only works started after the beginning of the depression, and they were never in good financial shape, judging from their subsequent changes in business organization and failure. They were probably near bankruptcy in 1877 and 1878 due to the continuing depression in the country. If they went bankrupt, their creditors would run the mill or sell it to others who would run it in order to earn the profits available with smaller capital costs. To make sense of the situation, it is necessary to assume

[18] AISA, *The Duty on Steel Rails*, pp. 13-24, 38, 62. See also *E&MJ*, 24 (1877), 301. Shinn left Edgar Thomson in 1879 with some bitterness. He apparently thought he could run the company independently of his majority stockholder, Andrew Carnegie. See letter of Carnegie to Shinn, April 4, 1879, Carnegie MSS, IV, and Bridge, pp. 124-130.

[19] U.S. Congressional Record, 47th Cong., 2nd Sess., 1883, p. 2295.

that the prime costs of manufacturing at the Vulcan plant were less than the selling price, and that there was some level of capital costs at which it would have been profitable to operate the firm. Otherwise the plant would have been unable to cover its prime costs and would have been shut down, leaving no motivation for the other steel mills to pay its debts. But if the costs were at a reasonable level, the other steel firms may have felt it worth while to capitalize on the situation by keeping the Vulcan Works out of production. The benefits from increased demand presumably more than offset the payment to the Vulcan Works.

The question that remains is why the steel companies did not let the Vulcan Works go bankrupt and then buy it to keep it idle, that is, bid up its price enough to keep others from buying it at a price that would make production profitable. A hypothetical reconstruction like the one above is not complete enough to answer this question, and the answer can probably be found only among the private papers of the individuals involved. However, we may support the above reconstruction and suggest that purchase was desired by appealing to the similarity of the Vulcan episode with the Joliet bankruptcy of 1877. In this failure, there was competition for control of the company between Meeker of Chicago and the other steel mills, represented by Morrell. It was asserted at the time that if the Eastern group had obtained control of Joliet, they would have closed it down.[20] The evidence for this is not strong, but it accords with the probable sequence of events at Vulcan.[21]

The collusion of the 1870's thus had some small effects in that decade, although the links between the actions just described and the profits of the steel industry are not clear. The main stimuli to profits were the general business revival and the railway boom at the end of the decade. Bridge reported that the profits of the Edgar Thomson Works were $1,625,000 in 1881, or 130 per cent of invested capital.[22] This is an extra-

[20] *Iron Age* (March 29, 1877), p. 11.

[21] It also conflicts in detail, but not in spirit, with the later assertion by the manager of Cambria that Holley, Hunt, and Jones wanted Cambria to restrict its output, but that he refused. John E. Fry, Letter to Sir Henry Bessemer, April 20, 1896, printed in *Scientific American Supplement, 41* (May 30, 1896), 17022.

[22] Bridge, p. 101. The profits of the Carnegie companies were over $2,000,000

ordinary level of profits, and it was a result of the small number of firms in the industry. The rapid rise in demand that caused the profit boom, however, in addition to the brevity of the boom, must be attributed primarily to forces outside the industry, and these factors restricted the growth of capacity as much as any arrangements of the steel men. No one, of course, had tried to enter the industry in the depressed years of the late 1870's.

In the course of the boom, many people saw the need for more capacity, and ten new Bessemer plants opened their doors in the years 1881-1885, of which half produced steel rails.[23] It was not in the interests of the Bessemer Association to have competing capacity, and they presumably tried to restrict the growth of competing plants. New firms were not admitted to the Association, and they did not have use of the patents still in force.[24] This placed them at a competitive disadvantage, but eight new companies were still making Bessemer steel at the beginning of 1886. The Bessemer Association was not able to control the growth of steel capacity, and a description of the new plants tells a little about why.

Five of the new plants were built by firms that did not make rails. They were not competing with the Association and were probably not bothered by it. Two of the new rail makers were protected from severe competition by their location, being in Colorado and Massachusetts. Another was the result of a split in the Lackawanna Steel Company, in which the manager and the brand name both went to a new firm bearing the name of both: the Scranton Steel Company. These were the eight new firms, and the Bessemer Association was not able to refuse them entry to the industry. The other two plants, however, were by 1886 part of the Bessemer Association. One was the famous South Chicago Works of the North Chicago Rolling Mill Company and represented expansion by that firm similar to the growth of other firms who did not have problems in their original site. The other was the plant at Homestead, Pennsylvania, built by a new firm in competition with the Edgar Thomson Works,

in 1881 and 1882, but they stayed below that level for the years 1883-1885. Bridge, p. 102.

[23] AISA, *Directory . . .*, 1890.

[24] U.S. Congressional Record, 47th Congress, 2nd Sess., 1883, p. 2257; *Iron Age* (March 13, 1884), p. 16.

and the only unsuccessful entrant. We ask if its failure was due to the policies of the Association.[25]

When the demand for rails rose in 1879, several Pittsburgh steel manufacturers had difficulty obtaining their steel from the Bessemer steelmakers, who asserted that their entire production was needed for their own orders for rails, and who refused to fill other orders. These Pittsburgh steel manufacturers consequently formed the Pittsburgh Steel Company, to build a plant at Homestead to make Bessemer steel for their own use. Since the mill was to make only billets, it would be substantially cheaper than a fully equipped rail mill.[26]

In 1883, despite the prosperity of the steel industry in the intervening years, the Homestead works were sold to Carnegie at cost, which was $1,200,000. A contemporary comment asserted that the "works have been admirably well built and very skillfully and economically managed." It was also noted that the mill had been converted to a rail mill in the process of construction.[27]

Two reasons are usually given for the failure of the Homestead company, to which we must here add a third. First, there was a strike when the mill opened over the question of its unionization, and the failure has been attributed to labor troubles.[28] The strike and the other labor troubles, however, appear much milder in the account of Fitch, a historian of the labor movement, than in the story told by Bridge, a historian of the steel industry. The influence of the strike appears to have been exaggerated in the accounts of men writing about the steel industry because it was a tangible event and could be easily linked to the subsequent sale. We may adopt the views of the labor historian and assign it a minor role in the developments which followed.

The second source of trouble for the Homestead firm was the decision to make rails rather than billets. This was an unfortunate decision for two reasons. The production of rails

[25] AISA, *Directory* . . ., 1890. The path of W. W. Scranton and the Scranton brand name may be seen in the *Directories* for 1880-1886, but the internal history of this movement is not clear.

[26] *E&MJ*, 28 (1879), 301; *Bulletin*, 13 (October 22, 1879), 267.

[27] *Bulletin*, 17 (October 17, 1883), 291; 17 (October 24 and 31, 1883), 297.

[28] Bridge, pp. 153-159; Casson, p. 111; Hendrick, Vol. I, p. 301; John A. Fitch, *The Steel Workers* (New York: Russell Sage Foundation, 1911), pp. 108-109.

required more capital than the production of billets, and the firm was unable to supply itself adequately for this task. In addition, the price of rails was exceedingly unstable. The decision to make rails was undertaken under the influence of the dramatic rise in rail prices in 1879-1880, and the decision to sell was taken under the influence of the equally sharp decline of 1882-1883.[29]

The switch to rails was an error of judgment on the part of the new firm, but there is no evidence that it was in itself a serious error. The Homestead firm, however, did not have the latitude to make even small mistakes. It labored under still a third difficulty: a cost disadvantage due to patent restrictions, and it needed more than merely capable administration to be a success. R. W. Hunt summarized the situation in the *Memorial* written for Holley in 1882. He was describing the value of Holley's patents, and he commented that only one Bessemer plant had been built in the United States that was "outside his [Holley's] patents." He then said, "I think my fellow Bessemer managers will unite with me in deciding it far below the standard of a Holley plant."[30] The Homestead plant not only had the labor troubles and insufficiency of capital that might beset any new firm, it had been denied the use of the patents that had made the American Bessemer industry a success, and its failure must be attributed at least partly to the restrictive policies of the Bessemer Association. This is recognized implicitly in discussions that treat Carnegie's opposition to the new firm, but it is never mentioned explicitly: the failure is seen as a lucky accident.

When Carnegie assumed ownership of the Homestead mill, he made extensive changes. The purpose of the mill was changed a second time, from rails to structural shapes. And those innovations which had been barred to the new company were incorporated into the mill. Carnegie later said that he had to spend $4,000,000 before the works "began to reach the turning point between failure and success."[31] This is certainly an exaggeration, but the need for extensive revision of the plant

---

[29] *Iron Age* (October 18, 1883), p. 14; Andrew Carnegie, *Autobiography of Andrew Carnegie* (Boston: Houghton Mifflin, 1920), p. 216. For monthly rail prices, see *AISA* (1904), p. 121.

[30] AIME, *Memorial . . .*, p. 31.

[31] *Bulletin*, *21* (December 7 and 14, 1887), 338; *32* (November 20, 1898), 177.

and its idleness for a few years while this was accomplished argues against Bridge's statement that "at the time of its purchase the Homestead mill was already one of the best-equipped plants of its size in the country."[32] The story shows the benefits of liquidity and of exclusive patent ownership.

Restriction of entry by means of patents helped the existing steel firms to exploit the boom of 1879-1882, and it helped refuse entry to at least one new firm in the following years. Several other new firms were successful, however, and we may conclude that the use of Holley's patents for restrictive purposes was not very important for the industry as a whole. In any case, these patents expired in 1886, and entry was free as far as patent costs went.[33] The large and ever-growing cost of an efficient Bessemer plant in conjunction with the high tariff continued to restrict entry into the American steel market, and profits to the steel firms were accordingly high. The argument so far indicates that an explanation of the level of profits is more likely to be found in the technical nature of production and the tariff than in the collusive actions of the steel firms.

The Third Decade

When the patents on the acid Bessemer process ran out, the men who had built the Homestead plant tried again to enter the field. Organized into the Allegheny Bessemer Steel Company, they opened a new plant at Duquesne in 1889, again close to and in competition with the Edgar Thomson Works. The Duquesne plant met the same fate as the Homestead plant; it was sold to Carnegie in 1890. In this case, Carnegie's opposition to the mill and the actions taken against it are openly acknowledged. The new firm was excluded from the rail pool, and railroads were advised against purchasing its product. The firm also had the labor troubles that plagued this particular group of men, and although they were less serious than the troubles Carnegie was having at the same time, they probably helped the firm in its decline.[34]

Nevertheless, this is the only plant for which a comparable

[32] Bridge, p. 161.
[33] *Bulletin*, 20 (February 10, 1886), 33; *Iron Age* (February 18, 1886), p. 18.
[34] *Bulletin*, 23 (February 20, 1889), 51; *Iron Age*, 46 (November 13, 1890), 854; Bridge, pp. 174-179; Casson, pp. 113-115; Hendrick, Vol. I, p. 366.

development took place after 1885, and eighteen new Bessemer plants were built in the years 1886-1889. There were also new plants using modified processes, such as the Clapp-Griffiths, Robert-Bessemer, and Tropenas. They were small-scale plants, designed to avoid the problems of the original large-scale Bessemer process, but they were never very important. And, finally, there was a growing number of open-hearth plants.[35] The Bessemer patents were not restrictive of entry after the early 1880's, and if there was a restriction by means of patents it had to be elsewhere. Chapter 6 showed that the open-hearth process was dependent on the introduction of the basic process for its success, and the basic patents were held by the Bessemer Association. The title was not clear, as there was a conflict of interests among inventors similar to the earlier conflict between Kelly and Bessemer, but the Bessemer Association apparently tried to restrict the use of the basic patents in the 1880's.

The basic process was discovered in the late 1870's, and patent controversies are said to have restricted its use in America for a decade after that. The contemporary accounts of this period, oddly enough, talk almost exclusively of the basic Bessemer process, ignoring the much more important application of the basic process to the open-hearth furnace. In view of the unsuitability of American ores for the basic Bessemer process, most of the discussion may be dismissed as irrelevant. With respect to the open-hearth process, however, the restricted availability of the patents retarded its introduction. The maximum length of time for which this delay may be attributed to the patent troubles is one decade, and the delay was actually substantially shorter than that due to the time necessary to learn about basic steel and to discover the possibility of using a basic lining in the open-hearth furnace. The delay did not have any permanent effects on the industry that can be detected, but to the extent that it restricted entry in the 1880's, it helped raise the profits of existing steel firms.[36]

The American rights to the basic process invented by Thomas

[35] AISA, *Directory . . .,* 1890, U.S. Census of Manufactures, 1905, p. 14. The census gives the number of Bessemer establishments as 11, 51, and 42 in 1880, 1890, and 1900. It gives the number of open-hearth establishments at these dates as 25, 58, and 96.

[36] Note that basic open-hearth steel was not made in England until 1885. J. C. Carr and W. Taplin, *History of the British Steel Industry* (Cambridge: Harvard University Press, 1962), pp. 102-103.

and Gilchrist in England were acquired by the Bessemer Association around 1880. They obtained the rights through the intervention or prodding of Carnegie, who in turn appears to have been prodded into action by that veteran of the industry, Abram Hewitt.[37] There was another set of patents, however, held by Jacob Reese of Pittsburgh. He, like Kelly before him, had anticipated the English inventors, but he had failed to make the new process a commercial success. He initially tried to introduce his process independently of the English process. Failing this, he assigned his patents to the Bessemer Company.[38] At this point the Bessemer Association either refused to grant licenses for the patents under reasonable conditions or else no one was interested in them. Either of these might be true; the distinction between them is not really clear. In any case, a controversy between Reese and the Bessemer Association started in 1885 and continued for several years. During this period, no one wanted licenses because there were no guarantees against later suits for patent infringement. The situation was similar to that with the Bessemer process before 1866, although unlike the earlier situation, this controversy was settled by the courts. The Supreme Court of Pennsylvania settled the controversy in 1888 by requiring Reese to give up certain of his patents to the Bessemer Company. The patents were not offered to the public, however, until after the Bessemer Association could reorganize itself into the Steel Patents Company in 1890. At that time they were offered for a royalty of $1 per ton of raw materials used, guaranteed against suits for infringement.[39]

There cannot have been much demand for the patents during this decade, because even the owners of the patents did not make much use of them. The Pennsylvania Steel Company started using basic linings in an old Bessemer plant in 1883,

[37] Hendrick, Vol. II, p. 39; Letter of George W. Maynard to Andrew Carnegie, April 8, 1892, Carnegie MSS, XV. A somewhat different account is found in Lillian Gilchrist Thompson, *Sidney Gilchrist Thomas: An Invention and its Consequences* (London: Faber and Faber, 1940), pp. 148, 160.

[38] *Bulletin, 13* (July 16, 1879), 183; *16* (August 2, 1882), 213.

[39] *Bulletin, 20* (February 3, 1886), 25; *21* (May 11, 1887), 124; *22* (October 3 and 10, 1888), 300; *23* (July 3, 1889), 177; *25* (January 21 and 28, 1891), 20. See also Victor S. Clark, *History of Manufactures in the United States* (3 vols.; New York, McGraw-Hill Book Company, Inc., 1929), Vol. II, pp. 268-269; *Iron Age, 46* (December 18, 1890), 1086.

but they abandoned the experiment in 1886.[40] And Carnegie instituted basic open-hearth steel manufacture at Homestead in 1883.[41] This was the only successful use of the basic open-hearth process in the United States before the patents were made available after 1890.

In this time also a few people not in the Bessemer Association tried the basic process. The Pottstown Iron Company produced basic Bessemer steel under a different set of patents, and the Otis Steel Company experimented with basic open-hearth steel in 1886. Basic production was said to be slower than that of acid steel, and the Otis Company changed back to acid steel.[42] This decision to abandon the basic open-hearth process does not accord with the calculations of Chapter 6, indicating either that the patent difficulties still apparent at that time influenced the decision, or that there was a larger discrepancy between Bessemer and open-hearth costs in 1886 than twenty years later.

Despite the controversy, therefore, the influence of patent control was not very important in the course of the 1880's. When conditions were unfavorable, or processes untried, people did not enter the industry. When conditions improved and new processes became known, many people entered the industry, most of them successfully. The timing of expansion may have been altered slightly by the patent arrangements, but there is little evidence of a major effect.

The failure of patent restrictions to retard entry does not mean that there was no collusion in the steel industry. As all collusion depends on restriction of entry, however, it means that collusion was more difficult. And as the heavy products typical of the early years of steel gave way to more varied objects, the limits on entry due to heavy capital costs declined for finished products. Capital costs continued to favor the creation and survival of pools in some areas, but these areas accounted for a decreasing part of steel production. Unless there was integration, there was no way to restrict entry in the lighter products, and collusion was not possible. Pools, however, were tried without integration, and only after this had been shown ineffective was integration increased.

[40] *Bulletin,* 17 (May 16, 1883), 132; 20 (February 10, 1886), 33.
[41] Hendrick, Vol. II, pp. 39-40.
[42] *Bulletin,* 23 (August 21, 1889), 228; *Iron Age,* 46 (December 18, 1890), 1086; *The Otis Steel Company* . . . .

The most successful pools of the late nineteenth century were the rail pools. The others led a fleeting existence without having much effect on the industry, and the influence of pools declined as the share of the market held by this heavy product declined. In other words, the effective concentration of the steel industry declined after 1880, although the concentration of the iron and steel industry rose due to the shift from iron to steel. Only after the advent of the holding company, a new device for market control, did concentration in steel production start to regain its old levels, and it is likely that the Bessemer patent monopoly and rail pool of the late 1870's represented the closest approach to a pure monopoly that the steel industry ever reached.

We are not interested in compiling a list of dates when pools were formed, changed rules, or dissolved. We want to chart their effectiveness over time according to some economic criterion. A pool ideally increases the revenue of its members by reducing output and increasing the price. The pool may therefore work by either restricting output or setting the price, letting the other variable find its own level. An attempt to do both can be effective only if the pool's forecasting of demand is exceedingly accurate; the attempt to do either can be effective only if it can be enforced. The most common method of enforcement was to require a deposit from the pool members which was available for fines and forfeit if the member withdrew. As the only loss from violating the pool agreement was the fine imposed or the loss of a not very large deposit, the pools were not permanent.[43]

The existence of stable prices over substantial periods of time is *prima facie* evidence of the existence of pools, although pools may exist without stable prices. The only product for which this evidence exists in the late nineteenth century is rails. There are two short periods of stable prices in 1875 and 1886, giving evidence of the pools of those years. The prices, however, were not maintained, and the effectiveness of these pools is not established by this means. Starting in February, 1891, the price of $30 per gross ton for steel rails was established for twenty-three months, after which the price was

[43] Wallace E. Belcher, "Industrial Pooling Agreements," *Quarterly Journal of Economics, 19* (November, 1904), 111-123. See also U.S. District Court (N.J.), United States *vs.* U.S. Steel *et al.* (1914), Government Exhibits, III, 975-983.

lowered in successive steps to $22 in 1895 and raised to $28 in 1896. These price changes were accomplished with never more than two months between successive prices, and all of the prices maintained lasted for six months or more.[44]

The pool of the 1890's had an effect on the price of steel rails that was more clearly seen than its predecessors of the 1880's. This reflects two things. The pools started their life by concentrating on limiting output only, and the effects of this action show up in high prices, but not in stable prices. By the 1890's, the steelmakers had decided that the division of the market was a less efficient means for maximizing their joint profits, or perhaps for dividing their joint profits, than the maintenance of a stable price. This, however, is not all that the prices reflect, for it appears that the pool of the 1890's was more effective than its predecessors, and that the stable price was a reflection of increasing power as well as a different form of control.

This may be seen through the movements of the ratio of the price of steel rails to the price of pig iron. The troughs of this ratio reflect the prices when the pools broke up and competition was allowed to operate; they therefore indicate the approximate costs of making steel in terms of the cost of pig iron. The peaks of the ratio indicate the level to which the relative price of steel was raised by the actions of the pools; they show how much this level exceeded the competitive price.

Since the cost structure of the industry changed rapidly in the 1870's, however, this method does not yield useful results about the effectiveness of pools in that decade. The ratio of prices declined, but it is not clear whether the decline was faster or slower than the decline in relative costs. It would appear that the combinations of the 1870's were not strong enough to overcome the price trend indicated by relative costs, although they may have retarded it.

The low point for the 1870's was about 2.2; the price of steel rails fell to 2.2 times the price of pig iron in 1877 and again in 1879. From this level the ratio rose to a peak of 2.7 in 1881,

[44] The criterion for "stable prices" used here is six months of the same price. The rail prices come from *AISA* (1904), p. 121. The pool of the mid-1880's was formed in 1885, but its duration is unclear. See *Bulletin, 19* (August 26, 1885), 224; *19* (December 16, 1885), 333; *20* (August 18 and 25, 1886), 220; *21* (November 23, 1887), 324.

reflecting the rise in demand of the boom and the lack of entry before 1881. The pool may or may not have been in active existence at this time; the important constraint was the one on output, not the one on price. This boom was the source of the fantastic profits of the steel companies that set the tone of discussion for many years, but it did not last. The ratio of prices had fallen to 1.8 by 1884, and the pool of the following years was only able to raise it to 2.1 by 1886, if indeed this small movement was due to the pool at all. For the period before 1890, therefore, we may say that pools as pools were not very important to the control of the market for steel rails, while the limitation of output due to natural barriers to entry, plus the tariff and patent control, enabled the rail mills to capture enormous windfall gains in the time of high demand. In times of low demand, a few of the existing firms were marginal, and it is not clear that further entry in the field would have lowered the profits to efficient companies like Carnegie's.[45]

The decade following 1890 was one of generally low demand and depressed conditions. Yet the ratio of prices we are discussing rose sharply and remained high until 1897. It was greater than two from 1890 to 1897, and it reached peaks of 2.5 in 1894 and 2.7 in 1896. This movement is remarkable as it is bracketed by values of the ratio of 1.9 in 1888-1889 and 1.7 in 1899, and it takes place in the course of depression. It could have been the result of a greater demand for steel than for pig iron in the depression, but this possibility is ruled out by the stable ratio of steel billet prices to pig iron prices in this decade. This ratio declined slightly over the decade, from 1.9 to about 1.7; it showed no hint of a rise in the demand for steel relative to the demand for pig iron in these years.[46] The

[45] The prices come from Appendix C, Table C.15. The various pig iron prices move together, and it does not alter the conclusions to substitute one for another. Gray forge pig iron is the one used for the 1880's, although No. 1 foundry pig is the only one available before 1873 and Bessemer pig is more suitable for the years after 1886 when it is available. Bridge, p. 102, gives Carnegie's profits for these years, but the amount of capital invested is not clear. The Union and the Vulcan works failed in 1883 and 1884. *Iron Age* (April 21, 1887), p. 33; *Bulletin, 18* (July 23, 1884), 188.

[46] The prices are still from Appendix C, Table C.15. The ratios used in the text are those involving gray forge pig iron. If Bessemer pig iron was used, the ratios would be shifted downward about 0.3, but the conclusions would not be changed.

rise in the relative price of steel rails for this period may therefore be attributed to the actions of the steel rail pool.

No other pool gives evidence of success as great as the rail pool. The production of beams would appear to have been equally well suited to pooling agreements, but the high investment needed to make beams was offset by the lack of uniformity of the product and smaller lots of sales. Although there were agreements maintaining the price of beams for many years in the late nineteenth century, there were no long periods of unchanging prices and no exceptional rises in the relative price of beams.[47] There were also pools and agreements for nails, plates, and billets, but these products were made by too many people and it was too easy for new people to make them for the agreements to be effective, at least in the nineteenth century.

The price of Lake Superior ore was set at meetings of ore producers, a practice dating at least from 1873.[48] As this variety of ore came to dominate the market, the price-setting actions of its producers became more effective. The constraint of competing production declined with the exhaustion of local ores in the older ore-producing districts. The Southern ores were not Bessemer ores, and imported ores represented only a limited form of competition to the Lake producers. The cost of transportation and especially the tariff ensured the domestic ore producers against an excess of price competition.[49] At the end of the century, however, the growth of non-Bessemer steel production increased the range of suitable ores and lowered the effectiveness of controls on the price of Bessemer ore. They did not reduce the effectiveness of controls on Lake ores in general.

The Formation of U.S. Steel

For products other than rails, then, the effectiveness of the pools was quite limited. Dissatisfaction with these relatively profitless agreements was a factor in the creation of the combi-

[47] U.S., Congress, Senate Report No. 2332, 58th Cong., 1st Sess., 1888, III, 975-977; Belcher, p. 119. A price series for angles, beams, and channels at Chicago is given in AISA (1904), p. 135.

[48] Bulletin, 7 (February 26, 1873), 205.

[49] Taussig, pp. 236-237, 299-300.

nations and holding companies that were incorporated into the
United States Steel Corporation at the turn of the century. The
creation of these new firms, however, was a complex phenomenon, and it cannot have been caused solely by the level of
profits. The best way to analyze this phenomenon is to jump
directly to the formation of U.S. Steel and to discuss other
firms in the course of discussing it. The new industrial giant
did several things. It provided a solution to a crisis in the
organization of the steel industry which had arisen in the 1890's,
it facilitated the formation and maintenance of various pools,
and it earned a large amount of money for the men who organized it. It was perhaps most successful in the last of these
roles in its early years, and this role may have been the one
uppermost in the minds of its creators.

The crisis in industrial organization came from the extensive
growth of integration in the last years of the nineteenth century.
Unlike earlier trends toward integration in the iron and steel
industry, this movement embraced horizontal as well as vertical
integration. The motives for vertical integration, which here
means ownership of ore and coal mines, not just blast furnaces,
appear to have been slightly different at the end of the nineteenth century than at the beginning. There was still the inducement stemming from the relative size of operations, the
minimum size of an economical blast furnace or steelmaking
plant being much larger than that of an ore mine or coke producers. But the job of organizing the mines or coke ovens into
larger concerns often had been done before the steel companies
joined the field, and the steel firms' motivations could not have
been the organization of this aspect of production. An additional motive was, however, supplied by the organization of
the ore and coke producing firms themselves, and by the concentrated geographical location of these raw materials. The
geographical concentration encouraged the growth of monopolies or monopolistic groups, and to the extent that the prices for
ore and coke reflected monopoly power, the steel companies
wanted to share in the profits. Having the capital to invest, they
became partners in this profitable field. The motives for combination were both the need for materials supplied on the technical basis required, as in the earlier period, and the desire to
share in monopoly profits, a motive peculiarly prevalent at the
end of the nineteenth century. The rise of the latter may be

attributed to the spirit of the age, but it was probably also due to the great increase in the scale of the efficient plant relative to the supply of raw materials.[50]

Horizontal integration was perhaps induced by the second of these motives, the desire for monopoly gain. "Trusts" were formed for many products, and the merger of many of them in U.S. Steel is an outstanding event of this period. With the exception of the two firms to be discussed next, almost all of these trusts were formed in 1898 or 1899 and they were almost immediately merged into U.S. Steel.

The conventional story of integration concentrates on developments in a few large firms. Carnegie merged his interests with those of Frick in order to produce his own coke and then with those of Oliver in order to mine his own ore.[51] Actions like this on the part of Carnegie and his competitors were a source of concern because they were linked to another development: the increasing power of a few corporations to disrupt the market by independent actions. We have already seen that Carnegie acquired the two mills that were built near Pittsburgh to compete with him. Possession of the three Pittsburgh plants made Carnegie an oligopolist whose actions had to be watched. Of approximately the same size and importance was the Illinois Steel Company, formed in 1889 by the merger of the three Chicago members of the Bessemer Association. The merger was essentially a purchase of the Joliet and Union Steel Companies by the North Chicago Rolling Mill Company, which thereby increased the number of its Bessemer plants from two to four.[52]

According to the conventional story, the great strength of these two firms, deriving both from their large size and their fully integrated status, promised a struggle to the death between them, and they were induced to combine to avoid the struggle. The result of the merger, U.S. Steel, relied heavily on these two firms for its output of steel, and this may explain

[50] This discussion of vertical integration, like the ones in Part I, has drawn freely on the discussion of Grosse. For the period discussed here, see R. N. Grosse, "Determinants of the Size of Iron and Steel Firms in the United States, 1820-1880," unpublished doctoral dissertation, Harvard University, 1948, pp. 287-288.

[51] These mergers are described in Bridge, pp. 167-173, 254-274, with all the appropriate drama.

[52] *Bulletin*, 23 (March 20, 1889), 76; 23 (March 27, 1889), 84; Clark, Vol. II, pp. 235-238.

why a firm was created containing them and only one other member of the old Bessemer Association. The new corporation, however, contained many more firms than these two, and it produced a wide range of finished products. The inclusion of the many recently merged firms making light steel products indicates that other motives for combination were present as well.[53]

One motive was undoubtedly to increase profits, and one way to do this was through pools. The pools appeared to work better after the turn of the century than before; the price of rails, for example, stayed at $28 a ton for several years after May, 1901. But the significance of these pools may be questioned. The ratio of the price of steel rails to the price of pig iron that had been above 2.0 for most of the 1890's was only above that level for one year (1904) in the first decade of the twentieth century.[54] This may indicate that the rail pool kept prices steady, but did not raise them, and therefore that the Steel Corporation did not distort the price structure on average.

A statement about rail pools is not a statement about profits in a fully integrated firm. The prices that U.S. Steel paid for its inputs were largely a matter of bookkeeping, and it could choose to take the profit at any stage in the manufacturing process. If part of the product was sold as raw materials or semi-manufactured goods, there was obviously an advantage in taking the profit at as early a stage as possible. The major significance of U.S. Steel in its early years, therefore, may have been the extent to which it facilitated the maintenance of high ore prices, for it did not do much to directly control production of its manifold products. Until the duties of the central office were expanded in the second decade of the century, they consisted of little more than the collection of uniform accounting information from the company's subsidiary firms. A thorough reorganization of the company's central office was not undertaken until the fourth decade of the century.[55]

Another motive for the formation of U.S. Steel, possibly the

[53] The story of the formation has been told often, and it exists in many versions. Clark, Vol. III, pp. 34-64, gives a moderate version and a good description of the background and results of the "great consolidation."

[54] Appendix C, Table C.15.

[55] Alfred D. Chandler, Jr., *Strategy and Structure: Chapters in the History of the Industrial Enterprise* (Cambridge: M.I.T. Press, 1962), pp. 333-334.

most important in light of the corporation's initial loose structure, was to profit from the act of formation itself. The profits from promotion were very large, and they were undoubtedly an incentive for action.[56] As they were dependent on the growth of a market for industrial securities—the profits came from selling the securities of highly capitalized new firms to an eager public—the mergers in the steel industry were part of a large merger movement at the turn of the century.[57] They may have been an attempt to profit from this movement, their effects on the American economy and in the steel industry being latent for many years after the merger. Questions such as this must await further study.

[56] U.S. Commissioner of Corporations, I, 167-179.

[57] Thomas R. Navin and Marian V. Sears, "The Rise of a Market for Industrial Securities, 1887-1902," *Business History Review*, 29 (1955), 105-138; Ralph L. Nelson, *Merger Movements in American Industry, 1895-1956* (Princeton: Princeton University Press (NBER), 1959), p. 37.

9

Changing Raw Materials

Iron Ore

The preceding chapters of Part II have chronicled the rise of the Bessemer and open-hearth steel producers. But these producers were not the entire iron and steel industry, although they represented an increasing proportion as the century went on. The proportion of the iron and steel industry's raw material inputs that were used to make steel was obviously minuscule in 1865; by the turn of the century, it was somewhat over three-fifths.[1] At any moment of time in these years, therefore, between 40 and 100 per cent of the iron and steel industry is unaccounted for. The purpose of this chapter and its successor is to describe these other portions of the industry, and to place the rise of the steel producers within the context of the industry as a whole.

Raw materials used by an industry, like other economic goods, are purchased or produced according to the supply and demand for them. Both the supply and demand for the raw materials of the iron and steel industry were changing their nature and their location at this time. On the demand side, the Bessemer process created a demand for iron ore low in phosphorus while hard driving in the blast furnace required a fuel that would burn rapidly and not be crushed. These were technical changes; the locational changes came from the expansion of the country

[1] Appendix C, Tables C.8, C.10. Almost three-fourths of all pig iron went to make wrought iron or steel in 1889, and four-fifths of the wrought iron or steel made at that date was steel. Multiplication shows three-fifths of the resources used to make pig iron going to make steel. In addition to this, there was the scrap and fuel used in the actual steelmaking process, which was only partially offset by the lower requirements for these materials in other branches of the industry.

194

that moved the consumers of the country ever farther from the center of the ante bellum iron industry: the "East." On the supply side, both the character and location of raw materials changed under the influence of changing costs at different ore and coal deposits.

The local iron ores of the East gave way to the Lake Superior, Southern, and foreign deposits as they wore out and as the need for Bessemer ores rose. They were too limited in extent and too high in phosphorus to supply the needs of the age. The anthracite deposits of the East gave way to the bituminous coal deposits in the West and South primarily for technical reasons; they too were not suited to the productive processes introduced by the new steel industry. We consider the changing ores and coal in turn.

The exploitation of the two new domestic iron ore deposits was the result of the expansion of the economy and the geographical exploration that accompanied it. This is obvious in the case of the Lake Superior ores, but it is no less true of the Southern deposits. The production of iron in Alabama in the 1880's was a closer cousin to the contemporary production in the North than it was to the ante bellum iron industry in the South.

The Lake Superior ore district, comprising deposits in Michigan, Wisconsin, and Minnesota, was discovered before the Civil War, but its exploitation began gradually. At first this was the result of transportation difficulties which were only partially solved by the construction of canals and railroads. The cheapening of costs had to wait upon developments in transportation that would increase the capacity of ore boats, facilitate mechanical unloading, and lower costs. And later, when the use of Lake ores became widespread, there was additional difficulty about their behavior in the blast furnace.

Lake ores accounted for about one-fourth of iron ore mined in the United States during the 1870's, and they rose after 1880 to one-half the total in 1890 and two-thirds in 1900.[2] It is surprising to find the relative expansion of Lake ores delayed in the 1870's, but there are reasons why the 1880's should have seen a rise. It was a period of great expansion for the industry in general, which would produce the need for new ore sources.

[2] *AISA* (1904), pp. 81-82.

And although we have seen in this decade the beginning of the great "battle" between Bessemer and open-hearth steel, the demand for Bessemer ore was still the most active part of the market. Lake ores were Bessemer ores in large part, and the distinction between them and Eastern ores was one of the reasons why Eastern ironmasters made foundry iron. The division between East and West comprised more than transportation costs and type of fuel; it extended to the type of ore available. The concentration of merchant blast furnaces in the East was not entirely a historical legacy from the new iron industry.[3]

The 1880's were also a time of mechanization in the industry and of railroad construction in the economy. Improvement in transport costs favored the Lake ores in their isolated location. Bell, on his 1874 visit to the United States, noted that the cost of assembling Lake ores and Connellsville coke at Pittsburgh was a dollar higher than the cost of assembling the mixed ores and anthracite in use in the Lehigh Valley.[4] One of the factors in the expansion of Lake ore production was the lessening of this disadvantage. A way to avoid it was to produce iron in Michigan, near the ore deposits themselves. This was tried, and Michigan was among the first five pig iron producing states in the 1870's. The proportion of the national production of pig iron made in that state, however, stayed below 5 per cent, while the proportion made in Pennsylvania stayed above 50.[5] The problem was that, although there was ore in Michigan, there was little else. Markets were lacking, and, more seriously, there was a lack of fuel, that is, of mineral coal. Of the locational influences in the nineteenth century, the proximity of fuel was one of the most important.[6] Fuel was not only bulkier than ore, it also deteriorated. The proximity of ore was therefore not sufficient to make the Lake Superior region a profitable one for the production of pig iron. Most of the iron produced in that region was charcoal iron and the course of technology ensured that this could not be a large proportion of the total. The new mineral fuel technology had made production with

[3] *Bulletin*, 30 (October 1, 1896), 218.

[4] Iron and Steel Institute (Great Britain), *The Iron and Steel Institute in America in 1890* (London, 1891), p. 48.

[5] Appendix C, Table C.12.

[6] See Walter Isard, "Some Locational Factors in the Iron and Steel Industry since the Early Nineteenth Century," *Journal of Political Economy*, 56 (June, 1948), 203-217.

charcoal too expensive for general use under almost any relative prices.[7]

Mechanization lowered the transportation costs of using Lake ores, both by increasing the facility of moving the ores and by increasing the ease of transshipping, loading, and unloading. The ore itself helped in this process by being a granular substance found virtually on the surface of the ground. The ore was scooped out of the ground by giant steam shovels, which were enabled to operate so crudely by the fineness of the ore and the isolated location of the mines. The giant steam shovels loaded the ore into railroad cars already joined into an ore train which was towed down a gradual incline to the shores of the Lake. It was then loaded onto a boat to be carried to Illinois, Ohio, or Pennsylvania, there to be unloaded at a steelworks or put onto trains for additional transport. The loading and unloading of railroad cars and ore boats was gradually mechanized, beginning in the 1880's. Ore was transferred from the railway to the boats by use of gravity, but the reverse procedure was obviously not subject to such a powering device. Hand unloading with powered lifts was used initially, but self-loading devices were introduced, and then buffer holding devices between boat and railway car further speeded the work. Speed was as important as direct cost in figuring the advantage of unloading docks as the Lake shipping season lasted only from May to November and the year's supply of ore had to be transported in that time. By the turn of the century the transport of Lake ores had become an intricate ballet of large and complex machines.[8]

Ore was taken to the blast furnaces, where it had to be stored for use during the winter. But the fineness of the ore that had made mining and transshipping so easy to mechanize made smelting difficult. Hard driving, it will be recalled, depended on the use of very high blast pressures to induce rapid combustion, and the fine ores had a tendency to be blown out of the top of the blast furnace rather than consumed by this pres-

[7] J. Stephen Jeans, (editor), *American Industrial Conditions and Competition* (London, 1902), pp. 401, 435; A. P. Swineford, *History and Review of the Copper, Iron, Silver, Slate and other Material Interests of the South Shore of Lake Superior* (Marquette, Michigan, 1876), p. 215; Chapter 10.

[8] Frank Popplewell, *Some Modern Conditions and Recent Developments in Iron and Steel Production in America* (Manchester: At the University Press, 1906), pp. 55-63.

sure. This difficulty was remedied, but the accumulation of ore dust in the waste gases and hot-blast stoves remained a problem for American blast furnaces.

The Southern ore deposits were mined by more familiar techniques, and their low cost derived from the proximity of pig iron production to the mines. There was coal in the South, and pig iron could consequently be made there with advantages. The proportion of American pig iron made in Alabama was close to 10 per cent in 1889, ahead of Illinois.[9] There would seem to be an inference that the locational effects were such at this time that the proximity of two of the three — ore, fuel, and markets — were needed to determine a favorable location for pig iron production.

The South contained ore and fuel before 1880, and most of the iron made was consumed out of the South even after that. How does one explain the late appearance of Southern pig iron in the national market? The answer would seem to be similar to the answer given in Chapter 3 to the question of why coke was not used thirty years earlier. The country was insufficiently known and the resources were not appreciated. The Southern iron industry in the slave era was ancillary to agriculture, and there does not seem to have been much incentive to look for better resources. The economy of the South and its small iron industry was destroyed by the war, and the chaotic Reconstruction period did not encourage the exploration of new opportunities. The often-heard plea that there was no market in the South may be interpreted to mean that there was no factor market, rather than that there was no product market, since the product was sold in the North after 1880. Once the Southern economy began to bestir itself in the 1880's, the ample resources for the production of pig iron were discovered and exploited, and "the Southern furnaces [became] the bugbear of the [Northern] trade."[10]

[9] Appendix C, Table C.12.

[10] *Iron Age* (July 7, 1887), p. 20. The standard source for this material is Ethel Armes, *The Story of Coal and Iron in Alabama* (Birmingham, Alabama: The Chamber of Commerce, 1910), but see also Lester J. Cappon, "Trend of the Southern Iron Industry under the Plantation System," *Journal of Economic and Business History*, 2 (February, 1930), 353-394; Robert Gregg, *Origin and Development of the Tennessee Coal, Iron and Railroad Company* (New York: The Newcomen Society, 1948).

A third new source of ore for the American industry cannot be attributed to geographical expansion of the American economy. These ores were imported from abroad, and their exploitation was a result of the demand for Bessemer ore. Whether or not they were known before, their rise to prominence was not the result of the entrance of a new area into the economic arena, and it was specific rather than general prospecting that discovered them if they were discovered at this time. Of the ores imported, well over half came from Spain in the 1880's, with French Africa running a rather poor second as an ore supplier. Sixty per cent of the ore was received at Philadelphia for the eastern Pennsylvania region, thirty-five per cent at Baltimore for the Pittsburgh area, and the balance at New York and New Jersey for local consumption.[11]

The geography of ore receipts shows what had happened to pig iron production in the North. It was no longer dependent on local ores by the end of the century, and there were very few local ores in use at that time. But the process was not merely that the Western producers used Lake ores and the Eastern used imported. The ore receipts show that imported ores were going to Pittsburgh, and the Lake ores were similarly going to the East.[12] Only the South and the nascent production in the far West lay outside this market for ore. The shift of location in the North, from East to West, was a small one in terms of distance, large though it may have been in symbolic importance, hardships on particular people, and difficulties of transport before the growth of a railroad network. Pennsylvania remained by far the largest producer of pig iron among the states, and Ohio was the second ranked state in 1840 as well as in 1900. There was a shift to be sure, but the unity of this area in the market for ore by the end of the century should restrain our emphasis on locational change. With the introduction of imported ores, there is no evidence that Eastern ironmasters paid a higher price for comparable grades of ore than Western producers.[13]

---

[11] *Iron Age* (March 10, 1887), p. 18.

[12] U.S., Congress, House Miscellaneous Document No. 176, 51st Cong., 1st Sess., 1890, p. 9; House Miscellaneous Document No. 43, 53rd Cong., 1st Sess., 1893, p. 263; House Document No. 338, 54th Cong., 2nd Sess., 1896, pp. 292-293.

[13] Appendix C, Table C.12; Appendix B.

Coal

The diffusion of coal supplies followed a pattern similar to that of ore supplies, with two significant differences. Due to the relatively high cost of transporting coal, the new coal supplies were more closely correlated with the new iron-producing regions. And the differences between the new coals and the old were slightly more subtle than the differences between the ores. The important chemical property of the ores was their proportion of phosphorus, the richness of the ore not being an important problem at this time. The only problem with the physical structure was to learn to use the fine Lake ores, a problem that was speedily solved. The chemical problem of new coal, as in the ante bellum era, was the proportion of sulphur. The Southern coal was sulphurous, and this probably helps explain why they were not used before the Civil War. In addition, they were not washed in the later part of the century, and the iron was impure. The famous low price of Southern iron must be at least partially balanced against its lower quality.[14]

Nevertheless, the discovery of a good coking coal was the key to the success of ironmaking in Alabama,[15] and there was good coking coal in plentiful supply around Pittsburgh as Connellsville became the standard source of coke for the Northern industry. The real start of Connellsville coke production had not come until the construction of the Clinton Furnace, the first blast furnace in Pittsburgh, in 1859, but the rise in the post-war years was rapid.[16] It was no accident that Frick moved to gain control of the coke fields in the depressed 1870's, at the same time that Carnegie was assuming control of the Edgar Thomson works.

It was the chemical properties of coke that retarded its introduction before the Civil War, but its physical properties assured its dominance as a blast-furnace fuel after the war. The process of coking was valuable for its effects on the sulphur content of the coal, but its main importance lay in its creation of a porous, yet sturdy physical structure. This was particularly true after the relatively sulphur-free, good coking coal of the Connellsville region was discovered and used.

[14] Iron and Steel Institute, p. 324.
[15] Armes, pp. 272-275.
[16] Howard N. Eavenson, "The Early History of the Pittsburgh Coal Bed," *Western Pennsylvania Historical Magazine*, 22 (September, 1939), 165-176.

Anthracite was a dense fuel, and although the lack of bituminous gas is common to anthracite and coke, the often-heard characterization of anthracite as a natural coke is not exact. For the lack of gas makes anthracite a very dense material, and the absence of gas makes coke a very porous one. When bituminous coal is coked, its volume does not fall. Rather than collapse upon itself, the remaining material in a good coking coal forms a fine honeycombed structure with air spaces in between. This means that coke has a higher surface area in relation to its volume than anthracite, and that it can burn faster. If a greater volume of air is blown into a blast furnace, coke will burn faster, while anthracite will speedily reach a limit set by the maximum rate of combustion on its restricted surface area. R. W. Raymond, president of the American Institute of Mining Engineers and editor of the *Engineering and Mining Journal*, told the members of the British Iron and Steel Institute at a joint meeting that if you use anthracite in your furnace, "after you give it a certain amount of air it will burn no more in a given unit of time in that particular air." He then explained why coke was added to the fuel in an anthracite furnace: "Our reasons for doing it is not because coke is a purer fuel, but because it will take more air and give the heat then and there, and anthracite will not."[17] Raymond was not saying anything new at this time; he was repeating what Eastern ironmasters had been saying for fifteen years, ever since Eastern furnaces began using coke as a result of a coal strike in 1875.[18]

When the users of anthracite had begun to mix this fuel with coke in the 1870's, they gave two reasons for their actions, beyond the temporary expedient of avoiding the strike. They argued on technical grounds, as mentioned just above, that coke was the better fuel. And they argued on economic grounds that the price of anthracite was too high. The two arguments are obviously connected, and the merits of anthracite were assessed at the prices reigning, but the arguments of the ironmasters

[17] *Iron Age, 47* (June 11, 1891), 1122. Raymond's statement is included in a report of the discussion at the joint meeting of the organizations mentioned. See *AIME, 19* (October, 1890), 932-995; *20* (June, 1891), 274-277.

[18] *Iron Age* (May 27, 1875), p. 17; (December 24, 1885), p. 7; (October 5, 1882), p. 23; (November 20, 1884), p. 11; *Bulletin, 11* (June 27, 1877), 170. Bell said that anthracite had the additional disadvantage of splintering in the furnace. See I. Lowthian Bell, *Notes of a Visit to Coal and Iron Mines and Ironworks in the United States* (second edition; Newcastle-on-Tyne, 1875), p. 35; *Iron Age* (May 15, 1884), p. 11; Iron and Steel Institute, p. 135.

illuminate why the price of anthracite was high. They talk in terms of willful exploitation of iron producers by railroads and coal combinations;[19] and they also talk in terms of competition for the use of anthracite. Anthracite was a smokeless fuel and a favorite for home consumption. This kept the price up, and although the price of anthracite did not rise relative to that of coke, it did not fall either. The troubles that stemmed from the small surface area of anthracite could not be avoided by using smaller sizes of anthracite; the coal companies' representatives stated in 1883 that "the demand for domestic [small] sizes of coal is so good that it would make no difference to the coal companies if all the blast furnaces served by them were to be thrown idle, as they could easily dispose of the furnace coal by running it through the breakers and putting it into shape for family use."[20] The two arguments of monopoly and competing demand are again connected, and it is a combination of both that kept the price of anthracite high.

Given these relative prices, we wish to see if the technical arguments reported above can be shown to have had a significant effect on the costs of anthracite blast furnaces. The price of anthracite at Eastern furnaces, despite the discussion, was lower than the price of coke at the same location, and if using coke was advantageous, it was for technical reasons. Can we discover the economic aspects of these reasons?

We were prevented from pursuing many questions about economies of scale and integration in steelmaking by the uncommunicative nature of the large steel companies. Iron men were in a much more competitive industry, and one which was not under attack at the end of the nineteenth century for its "exploitation" of the people. When Congress instructed the newly created Commissioner of Labor to ascertain the cost of producing dutiable articles, the Commissioner was able to compile much more information about the production of iron than of steel. He received cost data from over one hundred blast furnaces, but costs from only a handful of steelworks.[21] The latter are too few to be of use in a statistical investigation, and the information is not complete enough to allow a qualitative

[19] *Iron Age* (March 11, 1875), p. 11; (December 11, 1884), pp. 17, 25; *43* (April 4, 1889), 515.

[20] *Iron Age* (October 11, 1883), p. 14.

[21] U.S. Commissioner of Labor, 6th Annual Report, 1890, *Cost of Production: Iron, Steel, Coal, etc.* (Washington, 1891), pp. 31-179.

analysis. The former, however, provide an excellent statistical sample of the industry in 1890, and we may draw some conclusions from them. The statistical path from the data to the conclusions is drawn in Appendix B; the conclusions alone are given here.

It is found, by the use of statistical cost functions, that the use of anthracite was more expensive than the use of coke. Costs may be broken into material and nonmaterial cost as was done in Chapter 6 for steel costs. The nonmaterial costs for furnaces using anthracite were about one-third of a dollar higher than for those using coke. This difference may be attributed to the smaller size and capacity of the anthracite furnaces, that is, to the effects of hard driving on the cost of labor and related activities. The material costs were greater for furnaces using more anthracite than for those using less, about $.13 greater for each additional 10 per cent of fuel composed of anthracite. This result obtains despite the lower price for anthracite than for coke at Eastern furnaces; it is a result of divergent technology, and the contemporary observers who noticed the differences between the fuels were more accurate than those who complained of price discrimination. Taking into account the proportion of anthracite actually used by 1890, about 55 per cent in the furnaces sampled, the savings to be made at the location of existing anthracite furnaces by substituting coke for anthracite was over $1. The total cost of making pig iron was $12 or $13.

The effect on profit rates was greater than the effect on costs for two reasons. Profits are smaller than costs, and a decrease in costs means a proportionally larger increase in profits. (This assumes the selling price remains constant, but by 1890 all mineral fuel iron, with the exception of Southern iron, was sold together by grade.[22]) And the effects of hard driving, in addition to reducing the cost of a ton of iron, reduced the capital required for this production. The profit rate was therefore increased even further than the profit per ton by using coke, and anthracite can be said to have been definitely inferior to coke by 1890, perhaps not quite as inferior as charcoal was to it in the 1840's, but sufficiently inferior for the greater precision available in 1890 to detect it and measure its size.

We can also see the effects of location on the cost of coke

[22] See market reports in *Bulletin, E & MJ, Iron Age.*

and, as a bonus, the locational situation of anthracite furnaces if they had been using coke. It is found in Appendix B that the cost of coke varied with the distance, but that there ·was also a $.75 per ton price addition for anthracite furnaces. The use of anthracite may be taken as a proxy for Eastern, and we may conclude that there were more coke costs in the East than would have been anticipated from locational considerations alone, whether because the cost of transport was nonlinear, because this cost was higher in the East than in the West, or because of discrimination by the railroads against their captive blast furnaces. Nevertheless, this difference in coke costs was not added to a difference in ore costs since imported ore supplied the East. And it was balanced by the cost of transporting pig iron. In other words, blast furnaces in Eastern locations could compete in the East on the basis of a coke technology.

Anthracite blast furnaces, however, were not suitable for use with coke. By the 1890's almost all anthracite blast furnaces had some proportion of coke in their charge,[23] but there was a difference between a furnace using a mixture of coke and anthracite and one that used coke alone. The higher blast pressure necessary to burn anthracite coupled with the slow working of the furnace had produced an anthracite furnace design that differed sharply from the coke blast furnaces designed for hard driving. The experiments that were tried, willingly or not, with different fuels in different types of furnaces showed that a furnace designed for one fuel would not work well with another at the end of the nineteenth century. The typical blast furnace designed for anthracite or mixed coal was substantially smaller than the furnace designed for coke. If coke was put into the furnace, it burned more quickly than anthracite and failed to reduce the iron in the space and time allowed for it. The blast had to be lessened to allow the iron to be thoroughly transformed, and this lowered the output of the furnace. Furnaces could not be run profitably at low outputs, and experiments in this direction were not financially successful.[24]

[23] Appendix C, Table C.3.

[24] John Birkinbine, "Experiments with Charcoal, Coke, and Anthracite in the Pine Grove Furnace, Pa.," *AIME*, 8 (September, 1879), 168-177; Edgar S. Cook, "Anthracite and Coke, Separate and Mixed, in the Warwick Blast-Furnace," *AIME*, 17 (October, 1888), 124-129; Frank Firmstone, "Development in the Size and Shape of Blast-Furnaces in the Lehigh Valley, as Shown by the Furnaces of the Glendon Iron Works," *AIME*, 40 (February, 1909), 459-474.

The transition from anthracite to coke had to be accomplished, therefore, by the replacement of old furnaces by new ones. And the form this process took was the replacement of old firms by new ones. The people who came to the East to produce iron were the integrated steel companies, not the merchant blast furnaces. But their fortunes were not such as to encourage many firms to come to this region. The Pennsylvania Steel Company, the first Bessemer licensee, established a subsidiary, the Maryland Steel Company, at tidewater to use Cuban ores. The plant was started in 1887, but war and the panic intervened, and Lake Superior and Spanish ores had to be used.[25] Some difficulties on relying on foreign ores were not reflected in the price.

The Eastern region thus declined as a pig iron producing region, and the center of the industry shifted across the Alleghenies. The Eastern men, though, did not complain of their neighbors to the West nearly as much as they did of their Southern competitors. For although the West was growing faster than the East, it was the South that was coming to supply the Eastern market for foundry iron. Arising in the 1880's, the Southern iron was not pure enough to be used for the Bessemer process, but was good enough to be used in the foundry. Eastern ironmasters complained, but the Southern iron production grew.[26] Nevertheless, the production of iron in Alabama did not rise above the proportion of the national total reached in 1889. The Southern region was important as a marginal influence on the East, but it would not dominate the national scene; the center of the iron industry remained in the North, with the steel industry giants.

Smelting with anthracite may be seen, if we wish, as an analogue to the Bessemer process. Both represented the initial venture of a new technology; both produced a product having different characteristics than those known before; and both were abandoned in the course of technical progress. Smelting with anthracite took advantage of anthracite's purity and accessibility in the ante bellum era. As time went on, the importance of accessibility disappeared to be replaced by high costs based on deep mines. And as time went on, also, the purity of anthracite became of less value than the porosity of

[25] Harry Huse Campbell, *The Manufacture and Properties of Iron and Steel* (second edition; New York: The Engineering and Mining Journal, 1903), p. 679.
[26] *Iron Age* (November 27, 1884), p. 16.

coke. So too with the Bessemer process. This method of refining was based on the characteristics of a special kind of
iron ore and produced a product well adapted to a narrow range
of uses. When ways were found to use other kinds of iron to
produce a steel similar to Bessemer steel, and when this open-
hearth steel was found to be more adaptable than Bessemer
steel to products other than rails, the value of the Bessemer
process in America was sharply reduced.

We may even draw a parallel between smelting with charcoal
and refining by puddling. The analogy is less than exact, but
both processes were superseded by others because of their
inability to use energy in the form of heat. Charcoal was limited
by its physical structure which was even less susceptible than
that of anthracite to hard driving, and puddling did not employ
a furnace capable of generating sufficient heat to alter the
process. The products of these two processes, however, did not
disappear entirely in the nineteenth century. They were desired both for their qualities and for the small scale on which
they could be made, and they continued to be made for specialized or isolated small markets. This story is part of our final
chapter.

10

The Products of Iron and Steel

The Volume of Production

The iron producers of the ante bellum period carried over into the postwar years, and the new steel producers entered the industry with their revolutionary changes in methods of production, forms of organization, and type of product. We have examined the effects of this confrontation on the raw materials used by the iron and steel industry; we now ask about its effects on the products of iron and steel. We consider the volume of production and then its nature, taking the several different commodities in turn. The first discussion requires a consideration of the tariff; the others include a discussion of the materials produced.

The fluctuations in the volume of production were as severe in the years after 1865 as they had been before, but there was no parallel to the 1850's where a cyclical peak was not significantly larger than the previous one. These cycles have been very important for the economy and are easily seen in aggregate measures of manufacturing production. Frickey's index, for example, relies heavily on pig iron production and consumption. The iron and steel sector accounts for 20 per cent of his series, and pig iron consumed, a series derived from adjustments made on the production data, accounts for most of that. As this series is the most cyclically volatile of the important ones used, much of the cyclical pattern of manufacturing appears to be concentrated in iron and steel and their products.[1]

The important drops in production of the period were three

[1] Edwin Frickey, *Production in the United States, 1860-1914* (Cambridge: Harvard University Press, 1947), pp. 20-21, 158-169, 174.

207

in number, taking place in the middle 1870's, 1880's, and 1890's. The middle one was the mildest of the three, and the high growth rates for that decade are a reflection, or a cause, of this phenomenon.

The depression of the 1870's began with the crisis of 1873. It continued, in ever-worsening aspect, until about 1876 or 1877, when a trough appears to have been reached. Recovery came in 1879, with the steel industry responding to an expansion in the volume of railroad construction. The production of steel did not halt its upward movement during this depression, although the falling prices of these years brought forward several only half-remembered attempts at business combination, of which one was finally successful. The production of pig iron, however, fell by more than one-quarter of its previous peak before it began to recover.[2]

The boom that began around 1879 lasted for only a few years, and iron and steel firms began to complain of difficulties after 1883. This decline was not as severe as that of the previous decade, but it was more pervasive. The steel industry was growing too rapidly in the 1870's for the decline in business activity to affect its rate of growth, but by the 1880's the industry had matured to the point where the difficulties of that decade were reflected in a fall in the quantity of steel produced as well as in its price. The fall, however, lasted only one year, and the decline in pig iron output was smaller than in the 1870's, being less than one-eighth of the previous peak.

The 1890's saw business conditions somewhat comparable to those of the 1870's. Pig iron production fell about the same proportion of its previous peak and took several years to regain the previous high level. The low point of the depression in iron and steel was probably in 1894, with another relative low in 1896 after an intervening abortive recovery. The closing years of the century were the first part of a boom that fostered the merger movement of those years, which included the formation of the United States Steel Corporation.[3]

These cycles had a different relation to the international market than did the ante bellum fluctuations. The American ironmasters had been gradually increasing their ability to

[2] Appendix C, Table C.2. Estimated consumption of pig iron followed production closely [AISA (1904), pp. 96-97].
[3] Appendix C, Tables C.2, C.4.

compete for the American market, and they were helped by the Morrill Tariff of 1861 and subsequent revisions of the iron and steel duties. As a result of these duties and of the changing relationship of the American and British industries, the American iron and steel industry became the principal supplier of the American market, while British imports lost much of their importance. This had been foreshadowed in the boom of the 1850's when American rail production rose throughout the fluctuations in total demand, but it was not realized until after 1865. Thereafter, imports of iron and steel were quantitatively unimportant except during periods of exceptionally high demand. The boom of the late 1860's and early 1870's was accompanied by imports of pig iron equal to about one-tenth of domestic production and imports of rails equal to about one-half. The boom centering on 1881-1882 saw imports of pig iron of the same relative magnitude, but imports of rails equal to only about one-fifth of domestic production. From these cyclical peaks, the imports of iron and steel fell for the remainder of the century.[4]

The American industry was therefore freed from the British influence on cyclical movements, but it still relied on imports to fill domestic demand in heavy periods. Only toward the end of the nineteenth century was the American industry large enough to supply the home market without help in all conditions and to export its products when the home demand was low.[5]

The change in the relationship of the American iron and steel industry to its British counterpart at the time of the Civil War was a result of the tariff imposed during the war and of the gradually improving American technology. Both of these developments—the new technology and the new tariff—were at least partly the results of the iron and steel industry's own activities, and they were both promoted and supported in public discussion by the leaders of the industry. We have already seen the progress in technology and a little of the discussions in which American manufacturers defended hard driving and other innovations. Their support of the tariff was more vociferous, although it was marred by conflict within the industry.

---

[4] Appendix C, Table C.14, contains annual data on imports.

[5] The AISA collected export data, and they are scattered through their annual statistical reports, for example, *AISA* (1880), pp. 52-53; (1889), p. 60; (1898), p. 34.

The American Iron and Steel Association was one of the most articulate and obsessive of the tariff lobbies of this era. James M. Swank, who is the source of much of the information on the ante bellum era and much of the statistics of the period now under review, directed the AISA tariff lobby for many years, and we hear of his activities from many sources. Take for example the statement of a steel manufacturer appearing before a Congressional committee investigating tariffs in 1890: "I hardly know why I am here today, except we received notice in the public papers the committee wished to hear iron and steel men, and Mr. Swank, our representative, sent for us."[6] But Swank was not acting in a vacuum, and there is little reason to think that he was sufficiently charismatic to generate the protectionist sentiment with which he is identified. The nature of the protectionist paranoia may be seen in this comment of a blast-furnace owner in 1897: "I blew out on the 27th of August fearing Mr. Cleveland would be elected in 1892, and I was out twenty months. I blew out on the 2nd of September of this year for fear that Mr. Bryan would be elected, and I am going in on next Monday because Mr. McKinley was elected, and I am going to make 250 tons of pig iron every day."[7]

The AISA was identified with this wing of the iron and steel industry, but there were people on the other side of the fence. *Iron Age* went along with the *Bulletin* in its editorial policy, although it was not nearly as single-minded. The opposition journal was the *Engineering and Mining Journal,* which had a rather informal tie-up with the American Institute of Mining Engineers. In the course of the 1870's the *Bulletin* and the *E&MJ* conducted a running battle on the tariff, which presumably represents an antagonism between the AISA and the AIME. A characteristic incident resulted from a speech by R. W. Raymond, the editor of the *E&MJ* and a prominent name in the industry, at Liége in 1873. In this speech, Raymond invited I. L. Bell to visit the United States to see if the expansion of the preceding few years was going to lead to overcapacity and, according to the AISA, gave Bell and the British the idea that the United States steel industry favored free

[6] U.S., Congress, House Miscellaneous Document 176, 51st Congress, 1st Sess. (1890), p. 101.

[7] U.S., Congress, House Document 338, 54th Cong., 2nd Sess. (1896-1897), p. 319.

trade. This, of course, was not true and had to be refuted. The occasion was thus supplied for a reiteration of the positions of the respective journals. The *E&MJ* wanted to neglect protection and concentrate on improvements in the manufacturing process to help the American industry. The AISA was not convinced of the merits of the newest improvements in manufacturing, and it was sure that protection was the best way to prosperity.[8]

Not only did the AISA want duties on iron and steel, it wanted specific duties. *Ad valorem* duties were an object of condemnation before the AISA was formed,[9] and their condemnation on two separate grounds runs through the literature. Any duty that depends on the price of the article taxed needs a price of that article to be calculated. This gives an incentive to the importer to undervalue his imports to lighten his duty, which a specific duty does not, since the quantity of goods imported may be easily assessed. This was the motive behind classifying open-hearth steel as iron in its early years, and the prevalence of these actions made such free traders as David A. Wells support specific duties with the AISA.[10] The other reason was to make the duty protect more in depressions than in booms, rather than less. An *ad valorem* duty keeps the same percentage value over the cycle and falls in absolute amount when the price falls in depressions. A specific duty is a fixed absolute amount, and its rate rises when the price falls. Therefore an additional reason for specific duties was to insulate the American industry from fluctuations in the foreign price. This reason was undoubtedly the controlling one in the years before 1865, although the former reason appears stronger in later years.

A problem with specific duties, however, is precisely that they do not change when the price does. This may be fine over the cycle, but it creates problems when the price of a product,

---

[8] I. Lowthian Bell, *Notes of a Visit to Coal and Iron Mines and Ironworks in the United States* (second edition; Newcastle-on-Tyne, 1875), pp. 1-2; *E&MJ*, 17 (1874), 104-105; 22 (1876), 229, 233; *Bulletin*, 8 (October 15, 1874), 309.

[9] For example, Convention of Iron Workers, *Proceedings of a Convention of Iron Workers Held at Albany, N.Y., on the 12th day of December, 1849* (Albany, 1849), p. 28.

[10] U.S., Congress, House Executive Document 27, 41st Cong., 2nd Sess., "Report of the Special Commissioner of the Revenue" (D. A. Wells), 1869, p. cxxv; *Bulletin*, 4 (October 6, 1869), 33; 4 (December 22, 1869), 121. The AISA did not, of course, support Wells in general. See, for example, *Bulletin*, 4 (January 12, 1870), 145.

such as Bessemer steel in the 1870's and 1880's, is rapidly falling in what may be seen later as a secular trend. In this case, a specific duty becomes larger and larger as a percentage figure and more and more protective. The frequent adjustments of the tariff level in these years are more accurately seen as adjustments for price changes than as actual reductions in the rate.[11] An *ad valorem* rate might have been more suitable for this period when prices were changing in a secular fashion, while the specific duty would have been preferable for the period before 1865 when there were many *ad valorem* duties, and cyclical fluctuations were the major problems.

The actual effects of the high post-Civil War tariffs are somewhat problematical. At their best they worked just as theoretical discussion says they should. Take, for example, tin plate. All of the tin plate consumed in this country, with trivial exceptions, was imported from England before 1890. In 1890, the McKinley Tariff raised the duty on tin plate explicitly for the purpose of encouraging its domestic manufacture. If this manufacture was not forthcoming within a few years, the act stipulated that the duty was to be removed. The industry did take root behind its tariff walls, and the entire American iron and steel industry rejoiced.[12]

Rails are more typical and harder to discuss. The rate of growth of the production of rails in this country was not correlated with the level of the tariff,[13] and any simple test of the effectiveness of the tariff would probably show a similar absence of direct effects on domestic production. Yet Hewitt said that the difficulties of the steel industry in the 1880's were the result of overproduction in the 1870's caused by the high tariff rate. This was speedily denied by the AISA,[14] but it embodies sound reasoning. If the tariff increased profitability in the domestic industry, it thereby increased the incentive to expand production. Hewitt himself commented on the profitability of steel manufacture behind the tariff wall: "I have never known

[11] F. W. Taussig, *The Tariff History of the United States* (fourth edition; New York, 1898), pp. 221-224, 244-245.

[12] *Ibid.*, pp. 272-274; *AISA* (1911), I, 91-97; Victor S. Clark, *History of Manufactures in the United States* (3 vols.; New York: McGraw-Hill, 1929), Vol. II, pp. 372-374.

[13] Abraham Berglund and Philip G. Wright, *The Tariff on Iron and Steel* (Washington: The Brookings Institution, 1929), p. 116.

[14] *Bulletin, 18* (January 9, 1884), 10, 12.

of any such profits in connection with any business with which I have had anything to do." And again, "I suppose everybody admits that the duty upon steel is out of all reason."[15] These comments refer primarily to the extraordinary profits earned during the boom of the early 1880's, but industry profits were probably bolstered by the tariff in other, less spectacular, years also.

The tariff increased the incentive of American manufacturers to expand their production, although the extent of this increase cannot be known. The industry responded in the years following the extraordinary profits of the early 1880's, but it would be unwise to attribute the growth in capacity in the 1880's entirely to the effects of the tariff. And for other periods, even this much cannot be said.

If, as seems likely, the domestic supply curves for iron and steel products were either very elastic or else expanding rapidly enough to have the same effect, the effect of the tariff was primarily to replace imports by domestic production. The volume of domestic production, however, was still responsive to shifts in the demand for iron and steel products, and the effects of the tariff were masked. Only if the effects of the tariff were larger than the effects of shifts in domestic demand would simple tests reveal them. The only case in which this condition may have been met was the expansion of steel production in the 1880's.

By 1900, American steel men no longer considered the tariff a matter of importance. Carnegie, Gary, Wharton were all willing to abandon protection. As Wharton put it: "We are not pleading the baby act. We have grown to a stature where we can fight our own battles, and we merely demand of our legislators to give us a fair showing and let us in on equal terms."[16] This was not in fact a complete renunciation of protection as Wharton and others defined "equal terms" in such a way as to require some governmental help, but not as much as had been given in the past.

The volume of iron and steel production, then, increased rapidly — if a bit unevenly — in the years following the Civil

---

[15] U.S. Tariff Commission, *Report* (Washington, 1882), pp. 1085, 1087.

[16] U.S., Congress, House Miscellaneous Document 43, 53rd Cong., 1st Sess., 1893, p. 267; U.S., Congress, House Document 1505, 60th Cong., 2nd Sess., 1908-1909, p. 1748; *Bulletin,* 28 (January 15, 1894), 11.

War. This movement was aided to an uncertain extent by the tariff instituted during the war and maintained after its end, although the leaders of the industry were convinced that the tariff's effect was unimportant by 1900.[17] We now turn to an examination of the composition of this expanding production.

The Composition of Production

The proportion of pig iron that was used to make castings had reached its peak by 1850; its decline thereafter was slow and steady. From one-half of the total, the amount of pig iron that was not consumed in the manufacture of other forms of iron or of steel fell to one-fourth by 1900.[18] The consumption and production of iron castings, in other words, was not rising as fast as that of wrought-iron and steel products. The demand for cast-iron stoves, the largest single use for iron castings, did not rise as fast as the need for many other iron and steel products, and cast-iron pipe and machinery parts were being increasingly replaced by wrought-iron and steel products serving the same functions.

Despite this trend, a demand existed for charcoal pig iron in the late nineteenth century that was high enough to keep the differential in the price of coke and charcoal pig iron larger than the differential in costs for a small group of producers.[19] The charcoal blast furnace became larger, and it was reported that one made 1,000 tons in a week in 1890. This is questionable, considering the size of coke blast furnaces and the unsuitability of many innovations, for example, the Player hot-blast stoves, for charcoal; it is likely that the report was of a furnace mixing charcoal and coke.[20]

The demand for this specialized, high-priced iron came from several quarters. Car wheels, cast in the United States, were one large source. Malleable castings, castings with chilled surfaces, and castings requiring specially fine metal also de-

[17] See F. W. Taussig, *Some Aspects of the Tariff Question* (third edition; Cambridge: Harvard University Press, 1931), pp. 117-213, for an extended discussion of the tariff on iron and steel with conclusions only slightly less ambiguous than those articulated here.

[18] Appendix C, Table C.8.

[19] Iron and Steel Institute (Great Britain), *The Iron and Steel Institute in America in 1890* (London, 1891), p. 132.

[20] *Iron Age, 46* (October 16, 1890), 630; *Bulletin, 4* (November 17, 1869), 82.

manded charcoal pig. The growth of homogeneous steel production removed the demand for charcoal pig from rolling mills, but foundry demands remained strong enough to induce owners of wood distillation plants to produce iron as a by-product. In such a plant the furnace was less than 10 per cent of the total invested capital.[21]

Since foundries had become independent firms in the years before the Civil War, most pig iron not used to make wrought iron or steel was sold on the open market. The market classified charcoal pig iron separately from mineral fuel pig, but the latter class of irons was subdivided by grade rather than fuel. Foundry iron was sold for credit throughout most of the nineteenth century, but less and less credit as the century went on. Before the Civil War, six months' credit was the rule. During the war, iron manufacturers and dealers adopted new rules shortening credit from six to four months as of August 1, 1862. This period was eroded in turn by the growth of cash sales, and by the 1890's most pig iron was sold for cash.[22] By this time also, iron was sold f.o.b. at the blast furnace, the buyer absorbing the cost of transport as well as of finance.[23] As a result of these changes in market arrangements, the cost of pig iron to a purchaser was higher relative to the announced price at the end of the century than at the middle. Foundries presumably became more liquid, and the general availability of capital increased over this time, and this was why ironmasters could force their customers to alter the form in which they paid for their iron.

A further change was introduced into the selling of pig iron, but did not gain general acceptance in the nineteenth century. Pig iron warrants issued by warehousing firms were common in Britain, where they were part of the process of selling through a broker. The warrants provided a source of collateral for loans to blast furnaces, and the existence of pig iron stocks could even out price fluctuations. But these advantages were balanced by the dependence on centralized storage of iron with the

[21] Clark, Vol. II, pp. 250, 257; *Iron Age* (July 14, 1887), p. 14; Richard George Gottlob Moldenke, *Charcoal Iron* (Lime Rock, Connecticut: Salisbury Iron Corp., 1920), pp. 38-39, 49-54.

[22] *Hunt's*, 47 (1862), 186-187; market reports in *Bulletin*, *E&MJ*, *Iron Age*.

[23] *Iron Age*, 42 (November 1, 1888), 669. It is not clear whether this was a new development, like cash sales, or whether it had been true all along.

resultant effects on local monopolies stemming from transportation charges, and by an interference with the direct communication between buyer and seller desired by the former in a country with diverse sources of iron. Nevertheless, after several abortive attempts to operate a warrant scheme, the National Pig Iron Storage Association was formed in 1889 and achieved a gradual growth through the support of Southern iron men who sold in the North and were able to tap the credit sources of that region by the use of pig iron warrants.[24]

Although the proportion of pig iron that was used to make castings experienced a steady decline, the ratio of pig iron used in the manufacture of other forms of iron and steel to the output of these other forms, which was falling rapidly before 1865, did not continue to fall. The ratio of pig iron used to wrought iron produced continued to fall as scrap became an increasingly important input for iron rolling mills, reaching the low level of two-thirds by 1879. But the growth of Bessemer steel production acted against this trend. One of the important differences between wrought-iron and Bessemer steel manufacture, and one that encouraged the substitution of open-hearth steel for Bessemer, was that scrap could be used for only a small proportion of the Bessemer converter's inputs. The amount that could be used offset the waste involved in manufacturing the steel, but the ratio of pig iron used to steel products made was close to one. As a result, as more steel was produced relative to wrought iron, the ratio of pig iron used to rolled iron and steel products made rose to unity, where it stayed from 1890 till the First World War.[25]

From about 1830 to 1850, supply and demand forces were working in opposite directions as the supply curve for wrought iron fell faster than that for cast iron, while the demand curve for the latter rose quickly enough to offset the supply influence. By the Civil War, however, the demand for cast iron had ceased to rise enough to offset the supply influences, and the economy followed the dictates of changes in the supply curves for the various kinds of iron throughout the rest of the century.[26]

Puddled iron had been a new product of the new iron industry

[24] Clark, Vol. II, pp. 305-307; George H. Hull, *Reasons for the Establishment of a National Pig Iron Storage Association* (Louisville, Kentucky, 1888).
[25] Appendix C, Table C.8.
[26] See Chapter 2.

before the Civil War and had benefited from its falling relative price. In the post-Civil War era wrought iron was starving in the midst of plenty as the fall in the price of steel offset the increasing demand for iron and steel products and wrought iron was increasingly neglected. Despite its increasing relative price, however, wrought iron had at least three advantages over steel in the years following the Civil War. First, iron was easier to work with. It was less strong and could be shaped by smaller and less complex machines. *Iron Age* commented in 1883 that "iron has stubbornly refused to be forced out of use, and the indications are very favorable to the assumption that the country blacksmith, the carriage maker, the car builder, the machinist, the iron founder, and their multitudinous colaborers will continue to use iron for an indefinite period."[27] Second, wrought iron could be made from a wider variety of raw materials. Phosphorus was removed in the puddling process, but not in the original, acid Bessemer process. As was noted earlier, a large proportion of pig iron was thereby excluded from use in the Bessemer process. And third, Bessemer steel was subject to the mysterious breakages and fractures that made people prefer iron and later open-hearth steel in places where strain was a problem, and more susceptible to corrosion than iron which made people prefer iron where this was an issue.[28]

These difficulties were removed, one after the other, but their existence explains the slow demise of wrought iron and the ability of wrought iron to supply some demands better than steel for a long period after the introduction of the Bessemer process. The difficulty of working steel lessened in importance as people became used to steel and acquired the necessary heavy machinery. The problem of suitable raw materials was solved by the discovery of Thomas and Gilchrist in the late 1870's that by changing the lining of a furnace from an acid to a basic material phosphorus could be eliminated in the steelmaking process.[29] And the difficult qualities of steel were avoided in part by the substitution of open-hearth steel for Bessemer in the last years of the century.

The proportion of rolling mill products that were made of wrought iron fell steadily from its position of near 100 per

[27] Quoted in *Bulletin*, 17 (September 19, 1883), 257.
[28] Clark, Vol. II, p. 260; *Bulletin*, 19 (March 4, 1885), 59.
[29] See Chapter 6.

cent at the time of the Civil War to about one-fifth at the turn of the century. The proportion of the remaining production that was made of Bessemer steel stayed high for the last years of the nineteenth century, but one-third of the steel made at the end of the century was made in open-hearth furnaces. The proportion of steel production made in *acid* open-hearth furnaces, however, never rose above 10 per cent, showing the importance of basic linings for the change from Bessemer to open-hearth steel.[30] We now ask about the effects of these changes in the materials produced by rolling mills on the composition of their products.

Bessemer Steel and Rails

The changes in the composition of the iron and steel industry's output were a reaction to movements in both the supply and demand for these products. The revolutionary changes that were occurring in the production of steel during these years, however, indicate a profound realignment of the industry's supply curves, and the change in the materials produced may be attributed almost completely to this realignment.

There were no significant changes in the production of wrought iron in the years after the Civil War. The puddling process by definition could not use higher heat, for it would then be producing open-hearth steel; the attempts to mechanize puddling, such as the Danks revolving puddling furnace, were not very successful.[31] The changes in the production of steel, by contrast, were many and far-reaching; they have been extensively chronicled in the preceding chapters. The results were that the price of wrought iron in terms of other goods did not fall, although the price of steel fell sharply.[32]

The price movements of steel are very complex, due to the effects of changes in the general price level and to the many forms of market control in evidence at various times. Nevertheless, the ratio of the price of steel rails to the price of pig iron is a measure that can be used to give an index of the rate of change of the costs of making steel. Steel rails were a major product of the steel industry, and pig iron was its principal

[30] Appendix C, Tables C.5, C.10, C.11.
[31] Clark, Vol. II, pp. 259-263; *Bulletin*, 5 (September 21, 1870), 20; National Association of Iron Manufacturers, *Statistical Report for 1872* (Philadelphia, 1873), pp. 134-155.
[32] Appendix C, Table C.15.

input. The deflation thus produces a measure of the price a unit of the steel industry's output relative to the price of a unit of input. This ratio is not dependent on the general level of prices, and if taken at an appropriate time, it is free from the influence of monopoly pricing. The pools that fixed the price of steel rails did not last very long, and the price of rails in the intervals between pools was determined by competition among producers. The relationship between this competitive price and the level of costs prevailing at any time did not change over time, and we may use it as a measure of the industry's costs. The competitive eras in steel rail production are identified by looking at the low points of the ratio in question.

We find by this method that the relative price of steel fell throughout the last third of the nineteenth century, but at a declining rate. Before the Civil War, the price of steel was over five times that of pig iron.[33] In the years 1877-1879, when the depression of that decade was at its nadir, the price of steel rails was about 2.2 times the price of pig iron. In 1884, this ratio had fallen to about 1.8, and in 1899 — after the failure of the pool of the preceding years — it had fallen to 1.5.[34]

The price decline is evidence of technological progress; does the declining rate of descent indicate a slowing of this progress? Did the death of Holley in 1882[35] cause a change in the rate of technological change in the steel industry?

To answer these questions, it would be necessary to define a rate of technological progress more precisely than we have done here and to formulate explicitly a set of expectations on the "normal" rate of change of this rate over the life of an industry. The argument is often heard that it is easier to make innovations in a new field. If this is true, Holley may have been merely the reflection of this ease, and a different institutional form than the one adopted in the 1870's would have accomplished the same ends. But if it is false, a slowdown in the rate of progress in the Bessemer industry may have helped the fortunes of the open-hearth steel trade which challenged it in the years between 1880 and 1900.

In any case, the costs of making steel were declining sharply while the costs of making wrought iron were not. The in-

[33] George H. Thurston, *Pittsburgh as It Is* (Pittsburgh, 1857), p. 113.
[34] Appendix C, Table C.15. The prices are the same as those used in Chapter 8 to identify periods of collusion.
[35] American Institute of Mining Engineers, *Memorial of Alexander Lyman Holley* (New York, 1884), pp. 136-137.

novations that produced this fall pertained to the process of making the steel itself; they did not concern the making of individual products. There were innovations in the fabrication of many product lines, to be sure, but there were many such innovations, and none of them achieved the importance of the innovations in steelmaking itself. The costs of making any one product of steel fell just about as much as the cost of making any other.

The relationship between the prices and the costs of making iron and steel products also did not change substantially. The markets for products made of iron were competitive, as were the markets for most products made of steel. The product for which collusion was intermittently effective was steel rails. Even there, the increase in market control that appeared present in the 1890's and after the formation of U.S. Steel did not hold the price of steel rails permanently above what calculations on its ratio to the price of pig iron would lead us to expect would have prevailed in a competitive market.[36] We may conclude, therefore, that the supply curves for all products made of steel moved together, as did those for all products made of iron, and that the supply curves for steel products fell relative to the supply curves for products made of wrought iron.[37]

This argument and its conclusion about the movements of the supply curves resolves the identification problem implicit in a discussion of the product structure of the industry. Changes in the relative outputs that cannot be attributed to changes in the supply curve for steel may be taken to be the results of changes in demand.

But what changes in the product structure of the industry may be attributed to the changing supply curve for steel? The technical characteristics of steel differed from that of iron,

[36] See Chapter 8.

[37] A different version of essentially the same argument may be found in Peter Temin, "The Composition of Iron and Steel Products, 1869-1909," *Journal of Economic History*, 23 (December, 1963), 447-471. In that version, the starting point is the stable price structure of the iron and steel industry, considering the prices of the seven commodities listed in Appendix C, Table C.10. It is argued that the supply curves of the industry were elastic and that stable prices meant equal shifts in the relevant supply curves. The conclusion—that the supply curves for different products made of the same material did not shift relative to each other, but that supply curves for the two materials in question did—is the same.

The argument presented here has made use of Eugene Smolensky, "Comment" (on "The Composition . . ."), *Journal of Economic History,* 23 (December, 1963), 472-476.

and it was suited for different products. The composition of the output of steel differed from the composition of the output of wrought iron, and a shift in the relative amounts of the production of these two materials altered the composition of production of the industry taken as a whole. Changes in the product structure that were the results of the shift from wrought iron to steel, therefore, may be attributed to the changes in the supply curve for steel; changes in the product structure that were caused by shifts of either iron or steel production taken by themselves, on the other hand, must be attributed to shifts in the demand curves.

The products whose share of total output should have been raised due to the change in the supply curve for steel were rails and wire rods, plus structural shapes and plates and sheets after about 1890. The remaining products of the industry, bars and rods, skelp (strips used for the production of welded pipe), and nail plate, were more important in the production of wrought iron than of steel, and the falling supply curve for steel reduced their importance.

The actual changes in the composition of the iron and steel industry's output, however, were somewhat different. Rails fell in importance, rather than rising; and skelp rose in importance despite the contrary influence from the supply curve shifts. Nail plate fell in importance in accordance with the dictates of supply, but it fell more rapidly than would have been expected. Bars and rods fell also, but more slowly than would have been expected from supply considerations alone.

These deviations from the dictates of supply conditions were due to the influence of demand. We discuss the several products in turn, beginning with the most important product in 1865: rails.[38]

Steel was used almost exclusively for rails in the years of its initial great expansion, and rails were the first major product to be composed almost entirely of steel. This was accomplished in the 1870's, and when the price of steel rails fell below the price of iron rails in 1883, iron rails ceased to appear on the market except for specialized uses, such as light street rails.[39]

---

[38] The discussion in Temin, "The Composition . . .," again parallels this one, but from a more analytical viewpoint. The changes in the relative outputs of different products can be seen in Appendix C, Table C.10.

[39] Appendix C, Tables C.6, C.15.

Steel rails were made almost exclusively by the Bessemer process, although by 1900 the general switch to open-hearth steel had reached rails, and people were trying it, or complaining that the rail pool would not let them sell open-hearth rails at the premium they deserved, or that the effects of integration were to lock producers into the existing technology. The switch actually took place after 1900, and in the period we are considering the steel rail may be seen as the Bessemer rail without violence to the facts.[40]

After the price of steel rails fell below that of iron rails in 1883, there was little or no reason to use iron rails. Even before this, however, there was reason to switch to steel rails. The most important characteristic of rails was their length of life, and iron rails needed to be replaced often enough to make it worth a premium to have a longer-lasting rail. The *Journal of the Franklin Institute* did a comparative cost calculation in 1870 that shows the tenor of contemporary discussion. It said that iron rails lasted 4 years and steel rails, giving the reader a variety of alternative assumptions, 20, 40, or 60 years.[41] A commission appointed to investigate the life of steel rails in 1869 found that railroads reported having steel rails in good condition that had outlasted 13, 15, or 17 iron rails.[42] As a result of expectations and experiences such as these, steel rails were used for replacements in the 1870's, although new tracks were still laid with iron. This procedure enabled the lower priced iron rails to be used where they could stand up and the higher priced steel rails to be purchased only for those areas in which the high density of traffic or other factors necessitated a hardier rail. By 1877, presumably under this system, the Pennsylvania main line was entirely composed of steel rails.[43]

The demand for rails was a major inducement for the introduction of steel, and in the railroad boom of the early 1880's

[40] United States District Court (New Jersey), *United States vs. United States Steel Corporation et al.*, Government Exhibits, III, 1033, 1103; Harry Huse Campbell, *The Manufacture and Properties of Iron and Steel* (second edition; New York: Engineering and Mining Journal, 1903), pp. 530-531; Appendix C, Table C.11.

[41] Quoted in *Bulletin, 4* (March 9, 1870), 209.

[42] William H. Sellew, *Steel Rails* (New York: D. Van Nostrand Company, Inc., 1913), p. 4.

[43] *Bulletin, 11* (September 5, 1877), 237; George H. Burgess and Miles C. Kennedy, *Centennial History of the Pennsylvania Railroad* (Philadelphia: The Pennsylvania Railroad Co., 1949), p. 353.

the proportion of rolled steel products accounted for by rails was over 90 per cent. After that boom ended, in the recession of the middle 1880's and later, this proportion fell rapidly. It reached one-half about 1890 and had fallen below one-third by 1900.[44] Steel was moving out from rails in the years after 1880, into many diverse fields. It was in this expansion, rather than in the contest for rails, that the rivalry between Bessemer and open-hearth steel became acute. The uses of steel diversified because the increased size of the economy and the increased knowledge of these years increased the general demand for steel enough to compete with rails for steelmaking capacity, that is, because the demand for rails was falling relative to the other demands for steel.

Already in 1882, *Iron Age* had noticed that steel men were soon going to have to look outside the railroads for demand. By 1884 they could note that "If it were not for the demand for steel in other forms than rails the depression in the steel-rail trade would be severely felt, but there are very few works now dependent solely upon the demand for rails, and the tendency is to depend less and less upon this widely fluctuating business."[45] The fluctuating character of the demand for rails had been a well-known feature of the American scene, but it had not previously been central to the main part of the American rail-producing industry. In the ante bellum era, the boom demand was filled by the British industry and the American mills were able to rely upon the growing secular trend in rail consumption. In the early years of the steel industry the fluctuating parts of the industry were under the jurisdiction of the iron rail producers and the steel men had the rising steady demand that they liked so well. But in the 1880's, the capacity of steel rail mills had caught up with the trend in demand and the fluctuations in demand were transmitted directly to them.

The Bessemer steelworks began to produce other products, and new mills were increasingly built for purposes other than rails. But the Bessemer steel industry did not really survive the transition, although it diversified in the 1880's, as open-hearth steel was increasingly found more suitable for the uses that were then the source of the demand for steel. The Bessemer

[44] Appendix C, Table C.7.
[45] *Iron Age* (November 23, 1882), p. 14; (April 17, 1884), p. 19; (July 10, 1884), p. 16.

steel industry, like the puddled iron industry, was the response of the economy to the demands and supplies of the moment. They were both superseded as these conditions changed and the structure of costs and profits altered. Even though the steel industry of the nineteenth century presents to the superficial observer a record of unbroken rapid rise, underneath that rise are many smaller rises and falls of particular methods and business forms.

The *Iron Age* may again be allowed to survey the scene in this connection. It noted that the capacity of open-hearth steel plants in the 1880's was quite sufficient to supply the current demand for that kind of iron, but that new plants were being built rapidly to compete with Bessemer plants. Following hard on the heels of Bessemer steel's "victory" over wrought iron, indeed while the fight was still going on in most departments of the iron trade, a new front was opened by the open-hearth steel industry. "The record is one which may well cause the most sanguine to pause. It foreshadows changes in all departments of the iron trade. Some of them are now [1886] keenly affecting it, but it is not too much to say that they are merely the preliminary skirmishes of a great pitched battle."[46]

Open-Hearth Steel and Other Products

The battle between open-hearth and Bessemer steel was not fought on the plains of rail production. The costs of the former process declined at the same time that the demand for rails ceased to expand. The iron and steel industry was undergoing a transition in its product structure, and this was bound up with the changes in the materials produced.

Iron was being forced out of all uses by steel, although usually not nearly as rapidly as it had been forced out of rails. As the proportion of all products made of iron declined, iron became a product used only for a few products. By the turn of the century, 80 per cent of the wrought iron produced was used for bars, rods, welded pipe, and closely related products. These products were made from iron almost half of the time, while all other major classes of products were almost entirely steel. The proportion of iron to steel in these products was declining too, but more slowly than elsewhere.[47]

[46] *Iron Age* (June 3, 1886), p. 20.
[47] Appendix C, Table C.10.

Welded pipe was made from skelp, a form of flat bar made specially for that purpose. It was the competitor of cast-iron pipe, the type of pipe that had been important in the ante bellum years and that was experiencing a boom in the decade after 1883, as cities installed water and gas works.[48] Wrought iron was used in preference to steel to make welded pipe due to the difficulty of welding steel. This difficulty was overcome in the 1880's, and steel began to be used in increasing volume.[49]

Bars and rods not counted in one of the more specialized categories declined in importance over this period. They were the standard products of the ante bellum iron industry, but the increased supply of specialized steel products lessened their importance. Iron was still demanded here because of its ease of working, bars and rods being the product most widely disseminated to the small fabricator of metal products. The share of these products in total iron and steel production fell more slowly than would have been indicated by supply conditions alone as this demand remained strong. But the changes in supply dominated, and it did fall.

Their place was taken by a group of new products, one of the most striking and most often noticed of which was structural steel. Part I described the introduction of cast-iron and wrought-iron beams. After 1880, the production of structural shapes rose rapidly, but the production of structural iron stayed almost constant. The expansion was an expansion of structural steel.[50]

Structural shapes were used primarily to construct bridges and buildings, as in the ante bellum era, although other uses, such as ships, were present also. Wood, cast iron, and wrought iron were all used in construction in 1865. They continued to be widely employed despite the introduction of steel, but they declined in importance as the consumption of steel rose. Steel was occasionally employed in the form of steel rails, but this was not found to be a satisfactory arrangement, and structural shapes were produced as a separate item.[51]

These special shapes were a difficult product to manufacture. They had to be made in small lots as the structures varied among

---

[48] Henry Jeffers Noble, *History of the Cast Iron Pressure Pipe Industry in the United States of America* (New York: The Newcomen Society, 1940), p. 55.

[49] *Bulletin, 23* (May 8, 1889), 122; Clark, Vol. II, pp. 345-348.

[50] Appendix C, Table C.9.

[51] *Iron Age, 48* (July 2, 1891), 8; 52 (October 26, 1893), 758; *Bulletin, 17* (June 20, 1883), 165.

themselves to meet problems of location and terrain. And they became ever larger as the structures increased in size. Bridges became larger as engineers mastered the skills of spanning large distances, and buildings increased in height as the concept of a "skyscraper" based on a steel skeleton took hold after its introduction in the 1880's. The story was told in 1896 of the largest steel beam ever used in a building being hauled around New York by no less than twenty horses. Larger pieces had been used in bridges, but this 32-ton steel girder was the largest ever used in a building.[52]

In addition, open-hearth steel was found to be a more trustworthy material than Bessemer steel, and by 1899 less than one-third of structural steel was made from Bessemer steel, although all had been made from that material in 1879. When the steel manufacturers adopted specifications for their products in 1901, they allowed both kinds of steel to be used in buildings, but permitted only open-hearth steel to be used in structural steel for bridges and ships.[53]

Structural steel, however, was not the only new product that the steel mills were producing. The rapidly rising production of wire rods consumed an equal amount of steel and stimulated an equivalent expansion in the uses for which it was suitable. Wire had been made before steel became a mass-produced article, but the quantities had been small. Hewitt was able to claim that he made all the telegraph wire strung in 1850-1852, and wire fencing was not widely used because it broke under the continued pressure exerted by animals leaning against it.[54] Wire rope and other uses also consumed some iron, but the quantities were not large. We do not know precisely when wire rods began to expand as a proportion of rolled iron and steel produced, but when they began to be reported as a separate category around 1890, they were made almost completely of steel.[55]

The increased strength of steel and the growth of the econ-

[52] *Bulletin*, 22 (November 7 and 14, 1888), 330; *30* (October 10, 1896), 229; Clark, Vol. II, pp. 343-345; Edward C. Kirkland, *Industry Comes of Age: Business, Labor, and Public Policy, 1860-1897* (New York: Holt, Rinehart and Winston, 1961), pp. 257-258.

[53] Appendix C, Table C.11; Campbell, pp. 550-582.

[54] Allan Nevins, *Abram S. Hewitt* (New York: Harper and Brothers, 1935), p. 105; *Bulletin, 15* (November 9, 1881), 285.

[55] Appendix C, Table C.10.

omy increased the demand for the traditional wire products, but two new products were introduced in this period and probably accounted for the rapidity of the rise of wire. The first new product was barbed wire, which was invented in 1873 and which kept animals at a distance and saved the fence. The increased durability of a wire fence increased its usefulness while scarcity of wood on the prairies encouraged its use, and barbed wire was soon a common sight. The second new product was the wire nail, which replaced the traditional cut nail. Together these two new products accounted for something like one-half the consumption of wire in the 1890's.[56]

Wire nails were replacing cut nails for two reasons, their different shape and their different material. Steel was of increasing quality and cheapness; the price of steel nail plate probably fell below that of wrought-iron plate soon after the price of steel rails passed that of iron rails. And the round wire nails were found to be better than the square ones in addition, after some inventions of the 1880's, to being easier to make. They were widely criticized for their lack of holding power, but this was offset by their greater ability to penetrate wood without splitting it. They also weighed less than cut nails of equivalent length, which meant a larger number of nails in a pound and consequent lower freight charges. For these reasons, the production of cut nails reached a peak in 1886 and fell thereafter, and the production of wire rods for wire nails rose.[57]

The falling production of cut nails implied a falling production of plates for this purpose. The production of nail plate reached a peak in the 1880's, but it had been falling as a proportion of total rolled iron and steel since the Civil War. The relative fall in nail plate production meant a fall in the proportion of iron and steel used to make plates and sheets, although the proportion used to make plates and sheets for uses other than nails began to rise sharply after 1889 and regained for plates and sheets their former importance in the consumption of iron and steel. But, as in other products, the new uses for plates and sheets were best supplied by steel, and the proportion of iron used to make plates and sheets fell as the proportion

[56] Bulletin, 27 (November 22 and 29, 1893), 337; Clark, Vol. II, pp. 348-351.
[57] Iron Age (July 21, 1887), p. 17; (October 6, 1887), p. 15; Clark, Vol. II, pp. 351-355; AISA, (1904), pp. 113-115.

of total rolled iron and steel rose. And to complete the story, the steel that was used was open-hearth steel.[58]

The growing consumption of plates and sheets for other products than nails did not excite much comment as there was no spectacular product to be watched. Plates and sheets were used for boilers, ships, roofing, and a variety of miscellaneous products.[59] They were also used for one quite specialized use, which could not have used much steel although it generated a lot of excitement. This use was armor plate, and unlike the other types of plates, armor plate was a heavy product requiring heavy machinery to make. Entry was limited to the field of armor makers, and, in fact, the capital equipment necessary to make armor plate was so heavy and costly that no one wanted to go into the business. Only the government was in the market for this product, and if a company built an expensive plant to make it, it needed large-scale steady orders to realize a profit on the invested capital. In addition, since this situation created no incentive for more than one or two people to go into the business of making armor, a situation of bilateral monopoly was created between the government and the armor makers which gave rise to much acrimonious debate. It may be wondered that anyone went into such a business, but considerations of national pride and national defense made the government supply inducements to a few steel manufacturers to make armor.

Before the coming of steel, there had been armor, and part of the inducement to make a cheaper steel was to improve its quality. But this early armor was merely rolled iron plates; the specialized business of making armor came only with the start of the "new navy" in the 1880's, and the exploitation of new armor-making processes. The first contract in this series to an American company was with the Bethlehem Iron Company in 1887. It provided for an order of 6,700 tons of plain steel, oil-tempered and annealed, which was later modified to face-hardened (Harveyized) nickel steel. The difficulty of making this kind of steel and the cost of the plant required may be gauged from the contract price of $536 per ton, used even before the additional requirements were put on the steel. The Secretary of the Navy wanted an additional plant and, after some difficulty, persuaded the Carnegie Company to

[58] Appendix C, Tables C.9, C.10, C.11.
[59] *Iron Age*, 42 (November 8, 1888), 704; 44 (December 26, 1889), 993-996.

accept a contract. More contracts followed, the processes for making armor became more complex and presumably better, and Congress became uneasy about the price. There followed a long series of negotiations, some double-dealing, and probably some cheating on the contracts. But all that can be seen at this distance is the smoke of battle, not its precise outlines. For instance, the Carnegie Company apparently supplied defective plates to the government, or at least selected the plates to be tested to judge a lot on a decidedly nonrandom basis. When this was brought to light, the Carnegie Company asserted that all the plates were of good quality, and that they would replace any defective ones without cost. The Secretary of the Navy declined this offer and said he was sure every plate would satisfy the tests. It is no longer clear just what happened, much less who is at fault.[60]

Fortunately, the armor controversy was not of great importance, and we are not obligated to ferret out the secrets of this struggle. We are concerned here with the more prosaic movements in the consumption and production of all kinds of iron and steel. The "battle" between Bessemer and open-hearth steel is exciting enough for this history. Armor plates, and other plates and sheets, were made largely of open-hearth steel, and their increasing importance in the production of steel increased the importance of open-hearth steel. The Navy was interested in reliability, and the unexplained fractures of Bessemer steel made it undesirable. By the 1890's the Navy required open-hearth steel be used for armor.[61] For boilers and ship plates, similar considerations prevailed, and open-hearth steel was used. But we may not conclude from this observation solely that the rising demand for plates sparked the increasing production of open-hearth steel, for it was undoubtedly the case that the increasing availability of open-hearth steel spurred the consumption of plates.

This discussion has concentrated on heavy steel products, including the heaviest of all steel plates. The trend of the industry, however, was increasingly toward production of

[60] American Iron and Steel Association, *History of the Manufacture of Armor Plate for the United States Navy* (Philadelphia, 1899); *Bulletin, 21* (January 5 and 12, 1887), 3; *28* (July 11, 1894), 146; *32* (January 1, 1898), 2.

[61] U.S. Navy Department, *Circular Concerning Armor Plates and Appurtenances* (Washington, 1896), pp. 8-9.

lighter products, the production of sheets rather than plates. The steel industry's emphasis on the production of sheets in the twentieth century, and their growing importance in total steel production, may be attributed in large part to the demand from automobile production. This demand, which grew rapidly in the twentieth century, may be likened to the demand from railroads in the nineteenth century. It was large enough to direct the composition of production, and it sparked technological change in a particular, identifiable direction. By the beginning of the twentieth century, the rolling mill was the largest unit in the production of rolled steel products, and other units were grouped around it. The innovations in the production of flat steel products, such as continuous rolling mills, increased this trend and set the pattern of twentieth-century technological development.[62]

The iron and steel industry remained close to the demands of the transportation sector in the twentieth century, as it had done in the nineteenth. But the means of transport differed in the two eras, and the development of the iron and steel industry did likewise. This study ends with the end of the nineteenth century, and with the waning of its characteristic influences.

[62] See Douglas Alan Fisher, *The Epic of Steel* (New York: Harper and Row, 1963), pp. 137-151.

APPENDICES

Appendix A

Pig Iron Production Before 1860[1]

The purpose of this appendix is to examine the extant information about the volume of pig iron production in the ante bellum period. The data are fragmentary and of varying reliability; this discussion will provide an idea of the sources of the data and of the reliability of conclusions drawn from them.

The data on pig iron production, bad though they may be, are still far better than the data on the production of wrought iron or steel. Neither of these materials seemed as important to contemporary observers as did pig iron, and fewer attempts were made to collect information on the volumes of their production. As a result, it is not possible to give an idea of the annual fluctuations in their production. Information on the trends of wrought-iron production is given in Chapter 1; the production of steel in this period can only be listed as small.[2]

There are many different accounts of the volume of pig iron production before the Civil War, but they derive from a restricted range of sources. The five series shown in the first table of Appendix C comprise most of the contemporary estimates made of this production.

The series attributed to Henry Carey is the oldest of these tabulations, and it is the source for all later work on the 1840's. The American Iron and Steel Association (referred to henceforth as the AISA) accepted Carey's series, disagreeing only for 1840

[1] See Chapter 1.
[2] Steel was made from wrought iron and was very expensive. The demand for higher quality metal was not strong enough to call forth a large volume of production at the price at which people were willing to supply it. The price of steel at Pittsburgh in 1857 was over five times that of pig iron and over twice as much as that of wrought iron. (Implicit prices from George H. Thurston, *Pittsburgh as It Is* (Pittsburgh, 1857), p. 113.)

and adding a figure for 1820. The AISA collected much of the surviving data on the iron and steel industry of the nineteenth century,[3] and the series adopted by them is the standard series for the volume of pig iron production before the Civil War. It can be found in *Historical Statistics of the United States* in short (net) tons of 2,000 pounds with an accompanying note that attributes the statistics for 1854 and later years to the AISA, those for 1810, 1840, and 1850 to the Censuses of those years, and the rest to "early statisticians".[4]

The three remaining series were all compiled in the 1870's. They all take Carey's data as their starting point, but they treat them in varying fashion. Grosvenor was an ideological opponent of Carey; he wanted to support his arguments for free trade by showing that the production of pig iron in the United States did not rise rapidly during the high tariff 1840's, and he reworked Carey's data accordingly. Pearse presented the most complete set of figures he could find and included his sources with his table. His footnotes are not up to twentieth-century standards, but they provide more clues to the nature of his sources than the usual nineteenth-century omission of all notes. Raymond presented Carey's data with a few aberrations of his own, filling in the gaps by means of arithmetic interpolation.[5]

The individual figures have enough character to warrant examining them individually. We do so in chronological order, but most of the discussion will center on the problematical 1840's.

The figure for 1810 is the estimate derived by Tench Coxe for the Census of that year.[6] The collection of manufacturing data for 1810 was an afterthought, the main business of the Census being to collect population data, and the estimate must be judged accordingly. The scope of the Census grew over time at an uneven rate — manufacturing statistics were not gathered at all in 1830, and the Censuses of 1850, 1860, and 1870 were all gathered under procedures formulated in 1850. It was not until

[3] See the other tables of Appendix C.

[4] U.S. Bureau of the Census, *Historical Statistics of the United States, Colonial Times to 1957* (Washington, 1960), pp. 343, 366.

[5] R. W. Raymond, Appendix to A. S. Hewitt, *A Century of Mining and Metallurgy in the United States* (Philadelphia, 1876). Raymond does not say this, but an examination of the figures themselves, coupled with an absence of suitable original sources for, say, the 1830's, indicates this conclusion.

[6] Tench Coxe, "Digest of Manufactures, communicated to the Senate on the 5th of January, 1814," *American State Papers, Finance*, II, 666-812.

1880 that census data became anywhere near as complete or reliable as they are today.[7]

The figure for 1820 is of unknown origin; its initial appearance appears to have been in the work of B. F. French in 1858.[8] The block of figures surrounding 1830, by contrast, is of well-known origin. These figures are grounded on an actual enumeration of furnaces, and they are the first example of production statistics collected for the purpose of arguing about the tariff. The tariff in question was the Tariff of 1828, and the basis for the figures was a circular sent out by a free trade convention which met in Philadelphia in 1831. The replies to this circular gave information on the production of 73 blast furnaces and 132 forges for the years 1828, 1829, and 1830. The members of the convention knew of 129 other furnaces in operation at the time, and they credited them with a rate of production equal to the average rate of the 73 furnaces whose production was known. This technique of compensating for an incomplete sample will reappear several times in this account of the data.

A general convention of the Friends of Domestic Industry, an organization devoted to the maintenance of protection, met in New York in the same year as the Philadelphia convention. This convention thought the work of the earlier group was satisfactory and incorporated it into the report on iron. Before the report could be printed, however, a supplemental report was received, based on information collected by Peter Townsend. This was printed with the report, furnishing a revised estimate of 1830 production that was about 15 per cent larger than the original. The data were finally presented in three forms: the volume of pig iron production was given for the three years; the volume of pig iron production plus the pig iron equivalent of wrought iron made directly from the ore was also reported for all three years; and a corrected figure was given for 1830.[9]

[7] Carroll D. Wright, *The History and Growth of the United States Census* (Washington: U.S. Government Printing Office, 1900), pp. 20-29, 50-52.

[8] B. F. French, *Rise and Progress of the Iron Trade of the United States from 1621 to 1857* (New York, 1858), p. 20. French used this figure in reference to "several years" after the War of 1812. The Census of 1820 did not collect data on the quantity of iron production.

[9] Friends of Domestic Industry, *Report of the Committee on the Product and Manufacture of Iron and Steel of the General Convention of the Friends of Domestic Industry Assembled at New York, October 26, 1831* (Baltimore, 1832), pp. 13-16; French, pp. 32-48.

The fate of these figures is varied and confusing. Pearse was the only one who reported the figures correctly, giving the estimated volume of pig iron and direct castings produced and using the corrected figure for 1830. Everyone else used the figures for pig iron *plus* the pig equivalent of wrought iron made directly from the ore and the uncorrected figure for 1830. The figures shown for 1831 and 1832 should not have come from this convention as it was meeting in 1831, but it appears that at least one of them did. The 1831 figure seen in all columns is the corrected 1830 estimate of pig iron and pig iron equivalents produced, the initiative for this misreporting of the data apparently coming from Carey. Pearse, who should have known better, used the figure after citing Carey as its source. The 1832 figure appears to come from the same source, but is a straight extrapolation rather than a mistaken compilation.

Raymond's data for the 1830's are interpolations; Pearse's figures for 1837 and 1839 were taken from the work of J.H. Alexander.[10] The 1840 figure given by the AISA is the census figure for that year. The figure seen in Grosvenor's and again in Raymond's compilation is a correction of this figure by the Home League of New York, an organization favoring protection.[11] The figures used by Carey and Pearse are averages of the Census and Home League figures.

The Census of 1840 was universally acknowledged to be unsatisfactory by contemporary observers. The action of the Home League demonstrated a lack of respect for the census result by raising it some 20 per cent. Samuel Reeves, a member of the prominent Reeves family, commented that the Census of 1840 was "no credit to those who had immediate charge of it."[12] And, as shown in the table, most people did not use the census data. But whether the Home League estimate or an average of the two estimates is closer to the truth than the census figure must remain open to some doubt.

The data for the 1840's are the most controversial of the ante bellum period. They are not merely vague; they indicate startling movements that are of importance for the interpretation of the period. The peak production of 800,000 gross tons shown by

[10] J. H. Alexander, *Report on the Manufacture of Iron* (Baltimore, 1840), p. 68.
[11] *Hunt's, 12* (1845), 230.
[12] Samuel Reeves, Letter to Wm. M. Meredith, November 21, 1849, in U.S. Treasury, *Report on the Finances*, 1849, pp. 653-656.

Carey in 1847 is four times the production he shows for 1842 and over twice as large as even the Home League's estimate for 1840. There was a high tariff on iron in the years 1842 to 1846, and Carey attributed this rapid growth of production to the tariff. Grosvenor implicitly accepted his argument by trying to rework the figures to give the opposite conclusion. A more sensible approach is to examine the data a little more dispassionately, realizing that growth is a complex phenomenon into which the tariff enters as one element among many. Our approach here is to examine Carey's data and the later modifications of others. Carey's views on the tariff are considered in Chapter 1.

The data supplied by Carey for the years 1842 through 1849 appear to be his own creation, with the exception of the figure for 1846. The estimate for 1842 is an extrapolation from the data for 1840, and shows Carey's belief that the iron industry was seriously depressed. The data for 1847 through 1849 express Carey's opinion about the timing of the cyclical peak between the depressions at the ends of the decade.[13]

As extrapolations, these data are subject to some error. But if the basis for extrapolation is firm, the range given by the result may be relied upon. The depressed conditions of 1842 are not open to question; whether Carey's figure should be used as an expression of this is a matter of taste. The prosperity of the middle 1840's is also well known. But here we have a basis on which to evaluate Carey's work. We may trace the origin of Carey's one unoriginal figure, that for 1846, and we can test this figure against an independent compilation for Pennsylvania.

The figure of 765,000 tons of pig iron for the production of 1846 comes from a report of the Secretary of the Treasury, Robert J. Walker.[14] Walker derived his estimate from an estimate of Pennyslvania iron production by the Coal and Iron Association of that state. This estimate was reported in two slightly different versions by two of its members, and was derived in turn from the results of a partial survey of the industry made in 1842 by the same people.[15]

---

[13] Henry C. Carey, *Harmony of Interests*, printed in *Miscellaneous Works* (Philadelphia, 1872), p. 11.

[14] U.S. Congress, Senate, Executive Document 444, 29th Cong., 1st Sess., "Report of the Secretary of the Treasury, July 23, 1846," p. 91.

[15] C. G. Childs, *The Coal and Iron Trade, Embracing Statistics of Pennsylvania: a Series of Articles Published in the Philadelphia Commercial List, in 1847* (Philadelphia, 1847); Reeves, letter to Meredith.

The disreputable Census of 1840 was not felt to be a satisfactory starting point for an estimate of pig iron production in
1842, and the Pennsylvania Coal and Iron Association decided
to make a new start. They sent questionnaires to the blast
furnaces of their state requesting information on production. The
replies were disappointing: only 77 furnaces replied. Since there
were obviously more than 77 blast furnaces in Pennsylvania, the
number of replying furnaces was subtracted from the number
of Pennsylvania blast furnaces reported in the (unreliable)
1840 Census to get a number representing existing furnaces that
did not reply to the questionnaire. Production was attributed to
these furnaces at the average rate of replying furnaces (according
to Childs) or at a lower rate because of their inferior class
(according to Reeves). This procedure is the same as that used
by the 1831 investigators to complete their sample; the un-
certainty of extrapolation from incomplete data is shown by the
two different answers reported.

No new survey was conducted in 1846 to obtain the 1846
production. Production was imputed to 36 new anthracite
blast furnaces and 67 new charcoal furnaces that had been
built between 1842 and 1846 at an arbitrary rate (and, according
to Reeves, an allowance was included for the enlargement of the
furnaces existing in 1842). The two estimating procedures
yielded similar results for the production of 1846, but it was
clearly a hypothetical result. The procedures, for example, dealt
with changes in business conditions in an arbitrary way, ignoring
any differences in blast-furnace practice between the depressed
times of 1842 and the prosperous period around 1846.

The estimate of Pennsylvania pig iron production in 1846
presented by Childs (368,056 tons) was the starting point for
Walker's estimate of national production. Walker wrote in 1846,
and he dated his figures 1845. His source material and his later
followers all date their material 1846, however, and we may
follow their usage; the precise dating of the estimates depends
on assumptions about the timing of blast-furnace construction
which were not made explicit. To derive a figure for national
production from the data on Pennsylvania, Walker followed the
1842 practice of the Coal and Iron Association: he used the 1840
Census. He computed that iron production in Pennsylvania had
risen 274 per cent from 1840 to 1846. It seemed plausible that
the rest of the country had increased its production by 100 per

cent over the same period, and adding twice the census value
for production outside Pennsylvania in 1840 to the Coal and Iron
Association's figure for 1846 production in that state gave Walker
his estimate of national production.

Walker's final figure was the result of a highly ingenious chain
of estimation, but it still must be distinguished from the result of
an actual enumeration. An estimate is no more secure than the
assumptions on which it is based, and many of the assumptions
behind Walker's figure are of dubious merit. But, even a guess
may be accurate; we cannot conclude that Walker was wrong,
only that we are ignorant of the error in his result.

Fortunately, however, there exists an independent set of data
for Pennsylvania which permits a test of that part of the estimate
that refers to Pennsylvania. In 1849, a convention of ironmasters
met in Philadelphia to agitate for the reinstitution of the tariff of
1842-1846. They collected information on the blast furnaces of
their state, giving separately for each furnace its date of construction, largest attained product, capacity, kind of fuel used,
and other characteristics. The wealth of information presented,
together with its coherent organization, argues for its reliability:
the number of furnaces reported is close enough to that reported
by the 1850 Census to argue for the completeness of the convention's sample.[16]

Estimates of capacity are always dangerous to use. The convention of 1849 was aware of this and provided a report of the
largest product attained in the 1840's by each furnace in addition
to its estimated capacity. The largest product attained by the
furnaces constructed before the start of 1846 totalled 275,000
gross tons, of which 180,000 were made by blast furnaces using
charcoal. If the ironmasters were honest in reporting their
largest attained product, the annual rate of production at the
start of 1846 was equal to or less than 275,000 gross tons. This
figure is far lower than the 368,000 found by Childs for that date
and used by Walker.

What then was the peak production of the 1840's? Walker was

[16] Convention of Iron Masters, *Documents Relating to the Manufacture of Iron,
Published on Behalf of the Convention of Iron Masters which met in Philadelphia
on the 20th of December, 1849* (Philadelphia, 1850), Tables. The Convention
actually reported on more furnaces than the 1850 Census, but only about half
were in blast. The furnaces reported in blast by the convention were about 85
per cent of the furnaces reported by the Census; 150, compared with 180.
(Seventh Census, p. 181.) See also Chapter 3.

writing in 1846, before the end of the boom in furnace construction shown by the convention of 1849. He overestimated the production of furnaces that were producing at the time he was writing, but this does not say how far his estimate was from the actual peak of production reached. At the beginning of 1847, the cumulated largest product of the finished Pennsylvania furnaces had reached 341,000 tons, of which 208,000 were made with charcoal. But by the beginning of 1848, the total had only risen to 369,000; the boom in furnace construction was over by the start of 1847.[17]

It is probable that the production of existing furnaces started to fall off at the same time that the boom in furnace construction slackened. The peak production, accordingly, was not greater than the annual rate indicated by the furnaces in existence at the start of 1847: only if all blast furnaces were producing at their maximum rate in 1847 could such a rate be reached. This upper bound to the peak production of the 1840's needs to be corrected for the number of furnaces out of blast for reasons of economics and for repairs to given an accurate estimate of production. The magnitude of this correction is unknown, however, and we must be satisfied with the construction of this upper bound.

The method of extrapolation used by Walker to get a national figure from his data on Pennsylvania was clearly inadequate. To do better, it is necessary to use information that was not available to Walker, that is, information on the later development of iron production. Appendix C, Table C.12, shows that the proportion of United States pig iron made in Pennsylvania fluctuated around one-half for the last half of the nineteenth century. The Census of 1840 shows the proportion as one-third, significantly less than that for any of the later years. This figure may be due to the aberrations of the men in charge of the Census, but there is also a reasonable explanation for it, an explanation that provides a method of determining national production from that of Pennsylvania.

The 1840's witnessed a growth of pig iron production with mineral fuel. This fuel was used almost exclusively in the form of anthracite in the 1840's, the use of bituminous coal and coke being restricted to a very few furnaces.[18] The anthracite deposits

[17] All totals were computed from the Convention of Iron Masters, *Documents* . . ., Tables. The largest product of furnaces in existence at the start of 1840 was 160,000 tons, 139,000 which were made with charcoal.

[18] See Chapters 1 and 3.

of the United States are very highly concentrated in eastern Pennsylvania, and the production of pig iron with anthracite was concentrated in that state also. The production of pig iron with charcoal, conversely, was very widespread. If the Census of 1840 was accurate, it indicates that in 1840, one-third of the charcoal pig iron made in the United States was made in Pennsylvania, as all (or almost all) the production of that date was with charcoal. If it is assumed that the proportion of charcoal pig iron made in Pennsylvania remained constant during the 1840's, that is, if it is assumed that Pennsylvania contained one-third of the pig iron production made with charcoal and all of the pig iron production made with other fuels, the resultant share of Pennsylvania in national pig iron production that is extrapolated from the data of the 1849 convention agrees well with the Census started in that year. The rising proportion of mineral fuel production can account for the change in the proportion of production made in Pennsylvania and provide us with a method of extrapolating Pennsylvania production to national production.[19] Using this method, we arrive at an estimated volume of pig iron production in 1847 of about 750,000 tons, of which about 600,000 tons were made with charcoal. As indicated above, this estimate is an overestimate.

The conclusion of this discussion is that Walker was surprisingly accurate in his estimate of the peak rate of production in the 1840's, although that was not precisely his aim. His data, as used by Carey, overstate the volume of production according to the estimates constructed here, but the difference is not large if the above derivation indicates an actual production volume as well as a maximum possible volume.[20]

We have seen the source of Carey's figure for 1846; his figures for the remainder of the 1840's are extrapolations from it. It is probable that the peak volume of production was reached in 1847 and that the production in 1848 was lower than in 1847, but

[19] If the 1840 Census is not believed, the proportion of pig iron made in Pennsylvania could have remained close to one-half in the 1840's, the proportion of charcoal pig iron made in that state declining under the competition of the growing mineral fuel pig iron production. An assumption of this sort would yield a lower estimate of national production than the one in the text.

[20] Robert W. Fogel, "Railroads and American Economic Growth: Essays in Econometric History," unpublished doctoral dissertation, Johns Hopkins University, 1963, Chapter V, attempts to construct a continuous series of pig iron production in the 1840's to replace Carey's. I am indebted to Fogel for the opportunity to see his work and for discussion of the material presented both here and there.

the major outlines of the boom are as Carey shows them.[21] Grosvenor disagreed with Carey's view of the 1840's and attempted to prove that there was no boom between the lean years of 1842 and 1849 when the ironmasters were protesting their fate through conventions and memorials. His data show a rather even rise from 1840 to 1847, the peak production being 200,000 tons lower than that shown by Carey. Grosvenor constructed his figures from data available in the 1870's, starting from Carey's defense of his estimates for the 1840's in which he showed how his figures could be derived from the Census of 1840 and data on new furnace construction.[22] Grosvenor used the same data, but made different assumptions about their meaning: Carey expanded the 1840 volume of production to get an estimate of full capacity; Grosvenor assumed that capacity existing in 1840 was utilized to the same degree in all succeeding years as it was then. Carey assumed furnaces existing in 1840 were enlarged; Grosvenor assumed they were not. Carey assumed the capacity of new anthracite furnaces was 5,000 tons a year; Grosvenor assumed it was 3,000. Carey assumed that all furnaces built were used continuously; Grosvenor assumed that all furnaces that failed or were sold by the sheriff were kept out of blast permanently.[23] Neither set of assumptions is completely justifiable. As the above discussion indicated, the true value was probably between the two estimates. The production of pig iron doubled between 1840 and 1847, after which it declined temporarily. The precise quantity of production in 1847 is not known, nor is the precise path to it. It does seem likely that Carey was more correct than Grosvenor, however, in emphasizing the severity of the depression in 1842 and the consequent steepness of the rise to 1847.

The only new figure in Grosvenor's series is the one for 1845, which also appears in the remaining two series. It is another estimate of the Home League, and may refer to either 1844 or 1845.[24] The Census of 1850 was apparently respected by the observers of the iron industry and was used in all the series. The AISA series shown in Appendix C, Table C.2, starts in 1854; it is

[21] Reeves, letter to Meredith.

[22] Henry Carey, "Review of the Report of D. A. Wells," reprinted in *Miscellaneous Works*, p. 10.

[23] *Ibid.*; W. M. Grosvenor, *Does Protection Protect?* (New York, 1871), pp. 204-218.

[24] *Hunt's, 12* (1845), 231. The estimate is not dated.

shown in the AISA column of Table C.1. Raymond adopted these figures, but Pearse used an earlier version of them for 1854-1856 and reported them in net tons for 1857-1860 — a mistake in interpretation of the AISA's data.[25] Between 1850 and 1854 there were three years. Pearse took his figures for these years from B. F. French, who gives no sources.[26] Raymond appears to have taken his data from the same source and to have introduced a number of his own for 1853. The path of pig iron production in these three years remains obscure.

This completes the review of the ante bellum pig iron production data, with one exception. The Census of 1860 reported a figure for that year which was not used in any of the compilations shown in Table C.1. Unfortunately, the census figure does not agree with the data given by the AISA. The Census asserted that the production in the census year was about 1,000,000 tons; the AISA data point to a production level of about 800,000.[27] While this difference cannot be resolved at the present state of knowledge, it is possible to identify the source of the discrepancy.

An obvious explanation would be undercounting by the AISA, but unfortunately the Census only reported production in 1859-1860 from about half the number of establishments as the AISA cites in the mid-1850's. The two numbers of firms may not be inconsistent, due to the effects of the panic of 1857 and of technological progress in the course of the decade, but it would seem to exclude the possibility of undercounting by the AISA.

In addition, this creates another problem: to explain the discrepancy in per establishment production resulting from the larger production from fewer establishments reported

[25] The earlier version of the AISA data appear in J. P. Lesley, *The Iron Manufacturer's Guide to the Furnaces, Forges and Rolling Mills of the United States* (New York, 1859), p. 750.

[26] French, pp. 145, 150.

[27] There is some difficulty in comparing the two sets of data as the AISA data refer to calendar years while the census year stretched from June 1, 1859, to May 31, 1860. The problem has been solved by using approximate averages of the 1859 and 1860 AISA figures to compare with the Census. It is not necessary to be extremely accurate to show the discrepancy, and there is no evidence in the price movements or in the commercial journals of the time that there was a sufficiently extreme gyration of production to invalidate an approximation based on averaging.

There is an additional problem created by the fact that the 1860 Census did not specify the units it was using. It quotes the estimates of the previous Census, which everyone acknowledges was in gross tons, without indicating a change of units between the Censuses, and I have assumed the units were the same.

by the Census. This problem, however, has a solution. Pig iron was produced by different types of furnaces in this period, and the furnaces using mineral coal, whether anthracite or coke, embodied a new technology and were much larger than the older charcoal-using furnaces. A division of the census production between these various types of furnaces comes close to explaining the differences in per establishment production, although, of course, it cannot solve the problem of the aggregate totals.

The AISA showed, in round numbers, a production of 450,000 tons of iron with anthracite, 100,000 with coke, and 250,000 with charcoal, making a total of 800,000 for the census year 1860. The Census may be construed as saying the figures are 560,000, 175,000, 265,000, making a total of 1,000,000. The production of iron per establishment resulting from this distribution is approximately 7,500 tons for anthracite establishments, 6,000 for coke, and 1,500 for charcoal. Taking into consideration the fact that establishments using mineral fuel usually comprised more than one furnace, this agrees quite well with the AISA data, although the figure for charcoal firms is high.[28]

The primary discrepancy between the two sources is, then, in the production of pig iron with mineral fuel. They both agree that the production of iron with charcoal had declined by about half from 1847 to 1860, although there remains a possible pocket of disagreement over whether the number of charcoal-using firms fell even faster than the production of charcoal pig iron, as is suggested by the census data. In addition, the two sources agree reasonably well in their estimates of production per mineral fuel furnace and in the number of firms using mineral fuel. But the Census says that firms using mineral fuel were very

[28] The division of production was derived from county data. Production was assigned to charcoal or to mineral fuel furnaces according to whether the per establishment production was below or above 2,000 tons. For those counties in which there appeared to be a mix of furnaces, the production was split in such a way as to show the production per establishment to be close to 1,000 for charcoal firms and 4 to 5,000 for mineral fuel ones. A spot check of these results with the census returns, which have been collected by Henry Broude, shows them to be roughly accurate. There is a small bias tending to overstate the production of charcoal furnaces and to understate that of mineral coal ones, which derives from assigning production by deflating value, since charcoal pig iron was worth more than mineral fuel iron. This bias brings the census charcoal firm production closer to the AISA level of approximately 1,000 tons. The AISA listings of individual firms are given in the original compilation of their data, American Iron Association, *Bulletin*, and reproduced in Lesley. Lesley gathered the data for the Association.

active during and following the boom of the 1850's, while the AISA asserts the contrary by showing a smaller increase in the production of iron with mineral fuel in the years preceding 1860. To decide between these two descriptions, it will be necessary to find independent data on the establishments involved.[29]

I have used both sets of data, with allowances for their differences. Where a choice is necessary, I have relied on the AISA data as they are more complete, although I hope to have phrased my conclusions properly to avoid having them invalidated by further work on the data.

In summary, we may conclude that there is still a substantial amount of indeterminancy in our estimates of ante bellum pig iron production, but that the outlines are clear after 1840. The production of pig iron rose rapidly in the middle 1840's, reaching a peak around 1847 that was as high or almost as high as the production in any single year before the Civil War. Three-quarters of this production was made with charcoal, another fifth was made with anthracite, and the remainder was made with either bituminous coal or coke. The decade of the 1840's began and ended in depression, and the doubling of iron production in that decade was confined to a few years in its middle.

In the following decade, pig iron production rose from the level to which it had fallen in the depression following 1849, but it did not rise substantially higher than the level of 1847. But if the volume of pig iron production in 1860 was about the same size as that of 1847, its composition was different. Over half of the pig iron made in 1860 was made with anthracite, while only about one-third was made with charcoal. The proportion made with bituminous coal or coke rose above 10 per cent for the first time in 1860.[30]

[29] Grosse ("Determinants of the Size of Iron and Steel Firms in the United States, 1820-1880," unpublished doctoral dissertation, Harvard University, 1948) employs the original census returns for 1860. He does not reproduce the data, however, and his figures do not correspond directly with the published data, having too few firms in the states he considers and too many employees. He avoids the problems considered above by using employment as a measure of size; I was unable to find any complete statement about production per firm in 1860 in his thesis.

[30] The path of pig iron production after 1860 is given by the AISA series shown in Appendix C, Table C.2. These data agree well with the census data for the comparable years (shown in Appendix C, Table C.8), if conversions to gross tons are made for the census data of 1870-1890 (or 1869-1889, to use the more modern terminology). The latter two of these three Censuses specified that they were using net tons, but the first one did not. The Census of 1879, however, quotes the Census of 1869 as if it was in net tons, and that practice has been followed here too.

Appendix B

The Costs of Blast-Furnace Operations in 1890[1]

In order to see what factors were influential in the gradual abandonment of anthracite, several regressions were run on a set of cross-section data for 1889. The data were gathered by the Commissioner of Labor and consist of detailed cost accounts for 90 blast furnaces widely scattered in the United States, 14 of which used anthracite at this time.[2] The Census of 1890 reported that only 22.5 per cent of the blast furnaces in use in 1889 used anthracite, of which all but 3.4 per cent used a mixture of anthracite and coke for fuel.[3] Of the furnaces cited by the Commissioner of Labor, about 16 per cent used anthracite, but none of them used it alone. Since most of the furnaces using anthracite had already begun to mix it with coke, there are unfortunately no observations of furnaces using this fuel alone.

The sample of 90 furnaces represents somewhat more than 15 per cent of the furnaces in operation in 1889 as reported by the Census of 1890. The Census also reported that the average daily furnace capacity at that time was 68 tons. The average daily production of the furnaces in the sample was 92, but on the other hand the sample does not seem to include many of the largest furnaces in the industry, which were producing over 200 tons a day.[4] It seems fair, on the basis of the size and fuel distribution of the sample, to assume it represents conditions in the industry.

The sample can be divided into three subsamples: Northern blast furnaces producing Bessemer pig iron, Northern blast

[1] See Chapter 9.
[2] U.S. Commissioner of Labor, 6th Annual Report, 1890, *Cost of Production: Iron, Steel, Coal, etc.* (Washington, 1891), pp. 31-105.
[3] U.S. Census of Manufactures, 1905, Vol. IV, "Special Reports on Selected Industries" (Washington, 1908), p. 29.
[4] See Chapter 7.

246

furnaces producing non-Bessemer pig iron, and Southern blast furnaces (producing non-Bessemer pig iron). The furnaces in the sample show wide variations in the cost of making a ton of pig iron, with the subsamples appearing as roughly homogeneous groups having somewhat different costs than the other groups. As there is reason to believe that part of the differences in costs between the furnaces is related to the existence of these sub-populations, this variation must be corrected for before the data can be used. The difference between the subpopulations is primarily in their source of ore: the two Northern groups using ores from Lake Superior, but of different grades and costs, while the Southern group used Southern ore with a still different price. Putting the price of the ore used in the equations explaining costs might serve as a proxy for the different conditions under which the iron was produced.

Estimating an equation where the total cost of making a ton of pig iron, T, was made a function of the price of the iron ore used, I, resulted in the following equation, where the standard deviations of the estimates are given in parentheses:

$$T = 9.98 + 0.91\,I$$
$$(0.31)\ \ (0.07)$$

and R^2 equals 0.63 for this equation, showing that the price of iron ore used explained almost two-thirds of the variance of the total cost of producing pig iron. It is the final third of the variance that shall be our primary concern, but this result is significant in its own right. It shows that most of the variation in costs between furnaces producing iron of different grades and in different localities was the result of different costs of the ore they used. This is hardly surprising, but gives a sense of perspective to the discussion of relative costs. However, the coefficient of I is a bit unexpected. It is close to one, indicating that a rise in the price of ore of $1 would induce a rise in the total cost of production of about the same size. Since it takes close to two tons of ore to make a ton of pig iron, this means that the production function for pig iron was not of the fixed coefficient type. Whether the substitution reflected producer's reactions to changing prices or whether it was a reflection of high prices for richer ores is uncertain, but it indicates that there was an element of flexibility in iron production.

We first ask whether there was any inducement for the ironmaster in the anthracite region to substitute coke for anthracite in his blast furnace while staying in his location. Then we inquire whether there was an inducement for an efficient Eastern ironmaster to move elsewhere, that is, if he was at a competitive disadvantage and would be forced to move or retire from business eventually if he did not.

As noted above, after 1875 it was general practice to mix coke with anthracite in blast furnaces in the anthracite region. This could apparently be done without changing the form of the blast furnace or significantly altering the auxiliary equipment. The reasons given for this practice were that anthracite was troublesome to work in the blast furnace when it was alone and that it burned too slowly.[5]

Due to the increase in capacity of the furnace attendant upon using coke rather than anthracite, a reduction in the per unit value of those costs not immediately tied to the production of iron can be expected as these costs were allocated over a larger volume of production. On the other hand, this argument does not reveal any effect of changing fuels on the cost of materials necessary to produce a ton of pig iron.

To discover effects of different fuels, specifically of anthracite versus coke, on the cost of materials per ton on product, it might seem advisable to regress material costs, M, on the proportion of anthracite used. Neglecting for the moment the amount of raw bituminous coal used, which is not very significant in any case, an increase in this variable would represent a switch from coke to anthracite and its coefficient would measure the effect of this on total material costs. However, this variable could affect total material costs because of a technical factor or because the prices of the fuels, dependent perhaps on locational considerations, differed, their technical attributes in the blast furnace being the same. In order to separate these two effects, I define three variables as follows: Let A be the proportion of fuel used that was anthracite times the price of anthracite per ton. Let B and C be similarly defined for bituminous coal in its raw state and for coke respectively. By putting all three into a regression for M we obtain the desired information. For if the coefficients of all three variables are the same, then the effects of each on total material costs would be the same *if their prices were the same*. This regression can be seen as a price-corrected quantity regression, in

[5] See Chapter 9.

which the effects of altering the composition of the fuel input are seen independent of the prices of the various fuels. If the coefficients differ among themselves, then it will be true that changing the composition of the fuel affects the material costs even if prices are the same for all fuels and that this factor must be added to the change in cost resulting from any different prices.

Performing the actual regression gives the following equation, where I has also been included to account for the different subpopulations:

$$M = 5.10 + 1.03\,I + 0.64\,C + 1.34\,B + 1.20\,A$$
$$\quad\ (0.46)\ (0.05)\quad (0.12)\quad (0.50)\quad (0.15)$$

and R^2 equals 0.84 for this equation. The coefficient of B is not significantly different from either the coefficient of A or that of C, but these two coefficients are significantly different from each other.[6] There was a cost associated with using anthracite in a blast furnace in 1889, then, that was present even when the price of anthracite was no more than the price of coke. In other words anthracite was a costly fuel to use for technical reasons unrelated to location or market structure.

A dummy variable, equal to zero if a furnace was not using anthracite and one if it was, can be added to this equation to see if there was an addition to material costs incurred when a furnace used anthracite, but independent of the amount of anthracite used. Adding this dummy variable, D, into the equation yields the following result:

$$M = 5.14 + 1.03\,I + 0.62\,C + 1.33\,B + 0.86\,A + 0.73\,D$$
$$\quad\ (0.46)\ (0.05)\quad (0.12)\quad (0.50)\quad (0.28)\quad (0.52)$$

The coefficient of D is not significant in this regression, showing that the material costs of using anthracite were proportional to the amount used. The coefficient of A is changed very sharply, but not very meaningfully due to the high correlation (0.89) between A and D.

Turning to nonmaterial costs, N, there is no reason to expect the price of any fuel to be a factor in their determination. To see

---

[6] The test is to see whether their difference is significantly different from zero by means of a t test. The variance of the difference between two variables equals the sum of their variances minus twice their covariance. The value of the difference between the two coefficients is 0.56, and its standard deviation is 0.12. The t statistic is 4.7, and the difference of the two coefficients is significantly different from zero at the 5 per cent level.

whether nonmaterial costs were also sensitive to the presence of anthracite in the furnace, they have been regressed against the dummy variable for anthracite, D, and, of course, I:

$$N = 2.43 - 0.06\,I + 0.31\,D$$
$$(0.12)\ (0.03)\ \ (0.13)$$

and R^2 is only 0.11 for this equation despite the fact that all the coefficients are significant. Removing I from the equation lowers R^2 still further, but it does not significantly change the coefficient of D. We learn that the presence of anthracite did add about one-third of a dollar to nonmaterial costs, although the equation tells us that there were many other factors determining non-material costs.

In order to test whether the nonmaterial costs resulting from the introduction of anthracite depended only on whether anthra-cite was used or whether they also depended on how much was used a further test was run. Here N was regressed against I and a new dummy variable, D', equal to the proportion of the fuel that was anthracite. (D' is a continuous analogue of the discrete vari-able, D.) The results of this test showed D' not to be significant:

$$N = 2.46 - 0.06\,I + 0.36\,D'$$
$$(0.12)\ (0.03)\ \ (0.22)$$

The results are quite clear. The ironmaster who used anthra-cite in his blast furnace increased his costs over the costs of the ironmaster who did not. His nonmaterial costs were increased by the fact of his using any anthracite at all, while his material costs rose according to the amount of anthracite he used. How-ever, the magnitudes of these effects must be considered to assess their significance. Costs were higher by about $.30 for all anthracite users than for other blast furnaces. In addition, since they averaged 55 per cent of their fuel in the form of anthracite, there was an additional charge for this, mitigated, however, by the fact that the average price of anthracite at those furnaces using anthracite was about 20 per cent lower than the price of coke at these furnaces.[7] The "marginal" fuel cost per dollar of

---

[7] A t test may be run to make sure that 80 per cent of the coefficient of A is significantly different from the coefficient of C in the equation on page 249. The t statistic is 3.0, showing that this difference is significantly different from zero at the 5 per cent level.

coke used would have been $.64 had only coke been used. In the event, this cost was:

$$(0.64)(0.45) + (1.20)(0.80)(0.55) = 0.82$$

The difference is $.18 which must be multiplied by the average cost of coke ($4.15) to these furnaces to get the actual effect on costs. This makes the cost about $.75 per ton of iron produced. The extra cost from using anthracite thus seems to have been over $1 for those furnaces using anthracite, or close to 10 per cent of the total cost of $12 or $13. (The total costs shown in the report omit those costs included under general expenses in the earlier calculation. The omission may have been as much as $2.[8])

In order to convert this difference in the cost per ton into a difference in profit rates, it is necessary to consider capital costs, rates of output, and selling prices. Treating them in reverse order, we find ourselves immediately in a quandary. The above method was designed to use information relating to the manufacture of all grades of pig iron, each of which had its own price. By 1890 iron was sold by a classification system that defined, even if only approximately in many cases, the characteristics of the iron; the fuel used was no longer the determinant of price. However, the fuel used does not seem to have determined the grade produced, all types being produced with both coke and anthracite fuel. Therefore, although it does not appear meaningful to construct a composite price for pig iron, it can be assumed that the price of iron produced using competitively priced ore was competitive. In other words, there was not a compensating price differential to offset or partially offset the differences in costs as in 1850. With total cost being around $13, the profit margin must have been considerably smaller, and a $1 differential in cost must have represented a large percentage change in the profit margin per ton of iron produced.

It is to be noted that the argument here differs in emphasis from the argument about the earlier period in Chapter 3. There the cost data were poor while the output per unit of capital data were good. The cost calculation was done to show that the profit margin per ton of output did not decrease; the change in the capacity of a unit of capital was sufficient to show an increase in the profit rate of considerable proportions. Here the conditions

[8] *Iron Age*, 42 (November 1, 1888), 669; 46 (July 17, 1890), 94.

are reversed. The costs are known with some exactitude, while the cost of capital will be seen to be known only within fairly broad limits. The argument consists of demonstrating that the profit margin per ton of iron produced was greatly enhanced, while the output per unit of capital did not decrease.

The first part of this has been done. There was a definite saving to be made by reducing the proportion of anthracite charged into the blast furnace, keeping the location of the furnace fixed. This was the result of technological considerations that favored the use of coke, and which were only partially offset by a price differential in favor of anthracite at the locus of the manufacture of iron with that fuel. The effects of transportation will be dealt with later, but obviously do not enter into this argument. The savings to be made were considerable if one swung from an exclusive reliance on anthracite to an equally exclusive use of coke. The magnitude, calculated in a similar fashion to the immediately preceding calculation, gives a cost differential of $1.65, including the difference in nonmaterial costs. This is extrapolated from the above regressions on the assumption that the functions shown are in fact linear. Dropping the linearity assumption, the functions derived from the data can be used to approximate the real function near the observed points, showing there was a gain of over $.13 a ton of iron produced to be gained for every 10 per cent change in the proportion of anthracite used. This is known with a high degree of confidence, and it adds up to a considerable difference when large shifts are considered. And when compared with a profit margin per ton which could not have been larger than a dollar or two, the changes are seen to be very considerable indeed.

It is important to establish the connection between these profit margins and the rate of return on capital. If it can be shown that the output per unit of capital did not decrease when the blast-furnace fuel was switched to coke, the job will be done. It was mentioned above that a primary motive for changing to coke was in increase the output; in fact, the furnaces using anthracite in the sample used have significantly smaller outputs than the coke using furnaces, being about one-third smaller. Unless capital costs were more than one-third smaller for anthracite using furnaces than for furnaces using coke, output per unit of capital was increased by the change to coke. The evidence is imprecise; there are no direct cost comparisons. The impression gained from a survey of the field is that the equipment was slightly different for furnaces using the two fuels, but not appreciably more expensive for one or the other. While it was necessary to provide hardier equipment for anthracite users, the blast furnaces using anthracite tended to be older and smaller than those using coke.[9] This might make furnaces using anthracite slightly cheaper, but in view of the amount of rebuilding of furnaces that was done this difference could not have been as much as one-third.

A clear profit incentive existed, therefore, to reduce the amount of anthracite used in the blast furnace, lowering costs and increasing output at the same time. For the man using anthracite in the blast furnace, the switch involved no capital costs and the increase in output represented an increase in profit. This, however, was only true within limits and did not extend to the complete abandonment of anthracite. It appears that there was a discontinuity and that a furnace designed to use anthracite could work well on a mixture of widely varying proportions, but could not operate satisfactorily with only coke.[10] The argument presented, then, indicates why fuels were mixed in anthracite using furnaces, but does not say that anthracite users should have abandoned the use of that fuel altogether. This would have involved capital expenditure, and the data are not precise enough to say if the increased profit would cover this. But for a new entrant to the industry in this locality or for the ironmaster whose equipment needed replacing but who did not plan to leave the locality, it paid to erect new machinery appropriate to the use of coke rather than anthracite as a blast-furnace fuel.

But were there any such people? It would seem inconsistent to expect people to be rational with respect to technique and irrational with respect to location. Clearly if relative profits are to be used to explain behavior, they must be used consistently or be overridden by specific factors of greater importance where they are not used. It must be asked if there was an inducement to leave this locality.

The ironmaster using anthracite in 1889 was not at a cost disadvantage relative to other pig iron producers with respect to the

[9] John M. Hartman, "Notes on Fire-Brick Stoves for Blast Furnaces," *AIME*, 6 (1877-1878), 463-467; John Birkinbine, "The Distribution and Proportions of American Blast-Furnaces," *AIME*, 15 (February, 1887), 690-699.
[10] See Chapter 9.

cost of ore. The range of ore costs to anthracite-using furnaces was within the range of ore costs to all Northern furnaces, and if anything they were slightly lower for comparable grades of iron produced. The ores in use in the West came from the Lake Superior region, which presumably cost more to transport to the anthracite region than to the closer bituminous coal regions. However, the Eastern furnaces using anthracite did not use only this ore. They had access to local ore deposits and to the imported ore that was beginning to be major factor in the ore supplies of this country. Their ore costs therefore were not higher than the ore costs of other furnaces. To the extent that ore costs were the dominant locational influence at that time, as Isard asserts, furnaces in the anthracite region should have been under no disadvantage.

If we compare the costs of coke to various furnaces we find a different condition prevailing. The users of anthracite were paying significantly more for coke than were other furnaces, and furnaces using coke in the anthracite region must be presumed to have done likewise. If the price of coke P is regressed against the distance W from the coke ovens to the iron works and the dummy variable D, representing the use of anthracite, the following equation is obtained:

$$P = 2.59 + 0.0026\,W + 0.76\,D$$
$$(0.09)\ (0.0004)\quad (0.19)$$

and R^2 equals 0.54 for this equation which refers to the 81 furnaces of the sample for which the necessary data exist. The significance of the coefficient of D is that freight rates were higher in the East than elsewhere, giving rise to the hypothesis that the Eastern ironmasters were being squeezed out by the railroads. To discover how much the increased cost of these furnaces represented real costs is beyond the scope of this inquiry.

In order to find the total price disadvantage of the anthracite using furnace, W was regressed against D. This gave the following result with an R^2 of 0.15:

$$W = 122 + 176\,D$$
$$(19)\quad (48)$$

Combining the last two results shows that the furnaces using anthracite paid over \$1 more for their coke than other furnaces, about $.45 being accounted for by the greater distance from the coke ovens and $.75 being a result of other factors. The most interesting part of this conclusion is the relatively minor part that the distance consideration plays in it. To the extent that locational theory uses distance as a substitute for transport costs, it would seem to be omitting what may be the major part of a locational problem.

But before granting that the anthracite users were suffering under a cost disadvantage due to locational considerations, it is necessary to discover the costs of transporting the goods to the market. There was a growing Eastern market of considerable size which could support an Eastern blast-furnace industry if costs were right. It is to this market that we must look for the justification of the anthracite region pig iron producers as it was clearly more expensive to ship coke east and iron west than it was to ship neither.

Freight rates were an uncertain thing about 1890. In addition to the frequent changes, the practice of granting rebates made published rates often of little relevance. Nevertheless, they are all we have. A new set of freight rates in 1888 set the cost of transporting pig iron from Pittsburgh to Philadelphia as $2 a ton; From Pittsburgh to New York, at $2.40; from Pittsburgh to Baltimore, at $1.80.[11] An article the following year made the point that freight rates in the East were about $.01 per ton-mile,[12] which implies that the cost of transporting pig iron from a blast furnace in the anthracite region to Philadelphia or New York was in the vicinity of $.50. This means that the costs of transporting the iron to market from a furnace in the anthracite region were sufficiently lower than the costs from a furnace near Pittsburgh to balance the extra cost of coke at the Eastern furnace. The anthracite area cannot be said to be under a locational disadvantage with respect to the Eastern market relative to the other pig iron producers of the time. This does not say that this region was the ideal location for a blast furnace making pig iron with coke for the Eastern market. What it says is that an efficient furnace located there would not be undersold by other producers at their locations at that time. Any of these producers had the option of moving to more efficient locations, if they existed, which could make the anthracite region obsolete.

[11] *Iron Age*, *42*, (October 11, 1888), 539.
[12] *Iron Age*, *43*, (April 4, 1889), 515.

The argument is now complete. Anthracite was an inferior fuel to coke in the technology of 1890 for much the same reason that charcoal was inferior to anthracite forty years earlier. With neither of these fuels was it possible to get full usage out of the capital equipment, and both of them were abandoned in favor of more suitable fuels.

Appendix C

The Statistics of Iron and Steel

Sources for the Tables

Table C.1: *AISA* (1904), p. 91; Henry C. Carey, *Harmony of Interests*, printed in *Miscellaneous Works* (Philadelphia, 1872), p. 11; W. M. Grosvenor, *Does Protection Protect?* (New York, 1871), pp. 204-218; John B. Pearse, *A Concise History of Iron Manufacture of the American Colonies* (Philadelphia, 1876), p. 278; R. W. Raymond, Appendix to A. S. Hewitt, *A Century of Mining and Metallurgy in the United States* (Philadelphia, 1876).

Table C.2: *AISA* (1896), p. 60; (1887), p. 18; (1889), p. 30; (1892), p. 35; (1896), p. 34; (1901), p. 26; (1906), p. 42; (1911), I, 49. Mixed anthracite and coke production was not reported separately before 1883; it was included with anthracite.

Table C.3: Table C.2.

Table C.4: *AISA*, (1911), II, 57-61.

Table C.5: Table C.4.

Table C.6: Total rolled products, 1867-1886: Rolled iron plus rolled steel.

Total rolled products, 1887-1911: *AISA* (1911), II, 18.
Rolled iron, 1864-1890: *AISA* (1890), p. 50; (1896), p. 71.
Rolled iron, 1891-1903: Total rolled products minus rolled steel.
Rolled iron, 1904-1911: *AISA* (1911), II, 17.
Rolled steel, 1867-1885: Regression result—see next page.
Rolled steel, 1886-1890: *AISA* (1887), p. 43; (1896), p. 72.
Rolled steel, 1891-1903: Regression result—see next page.

Rolled steel, 1904-1911: *AISA* (1911), II, 17.

Total rails, 1867-1904: *AISA* (1904), p. 102.

Total rails, 1905-1911: Iron rails plus steel rails.

Iron rails, 1849-1911: *AISA* (1911), I, 86.

Steel rails, 1867-1911: *AISA* (1904), p. 102; (1911), II, 60.

To get the quantity of rolled steel products for years in which it was not reported, the quantity of rolled steel products was regressed on the quantity of steel ingots and castings (Table C.4) for years in which it was reported. The regression result was:

$$S = (0.8426 - 0.0022t)\, I$$
$$(0.0534)\ (0.0013)$$

where S is rolled steel, I is ingots and castings, and the standard errors are shown below the estimates. (R^2 for the regression was 0.99.)

Table C.7: Table C.6.

Table C.8: 1849: Seventh Census, "Compendium," pp. 181-82.

1859: Eighth Census, III, clxxx, clxxxiii, cxciv.

1869: U.S. Census, Ninth, Vol. III, "Wealth and Industry" (Washington, 1865), pp. 603-607, 625; U.S. Census, Tenth, Vol. II, "Manufactures" (Washington, 1883), p. 738.

1879: Tenth Census, II, 738, 743.

1889: U.S. Census, Eleventh, Vol. VI, "Manufacturing Industries," Part III, "Selected Industries" (Washington, 1895), pp. 420-422. U.S. Census, Twelfth, Vol. X, "Special Reports on Selected Industries" (Washington, 1902), pp. 36-58.

1899: Twelfth Census, X, 36-58, 74.

1909: U.S. Census, Thirteenth, Vol. X, "Reports for Principal Industries" (Washington, 1913), pp. 217-245.

Table C.9: 1869: Ninth Census, III, 604-625.

1879: Tenth Census, II, 738-739.

1889: Eleventh Census, VI, Part III, 420.

1899: Twelfth Census, X, 59.

1909: Thirteenth Census, X, 238.

Table C.9 reproduces the data contained in the Censuses for 1869 through 1909 on the volume of production of rolling mills and steelworks, of which all but a very small proportion were rolled products. No efforts were made to fill all the gaps left by the Census, the only changes made being those necessary to preserve a consistent system of units and classification. The data for 1869, 1879, and 1889 were converted to gross tons to agree with the later years. The total quantity of rolled iron products for 1869 was not given in the Census and was estimated by assuming that the share of "other products" in the quantity of production was the same as its share in the value, which was given. And the figures for 1889 were corrected to eliminate double counting deriving from the inclusion of rolled products sold within the industry for further manufacturing in the category of "other products" and, consequently, in the total production of finished goods by the industry. (The note in the Census that said this was appended only to the data for iron, but the discussion in the text obviously referred to steel. In addition, the magnitude of "other product" production – close to 30 per cent of the total – shown by the uncorrected Census data was unrealistic.)

Table C.10: Table C.9 and *AISA*.

Table C.10 differs from its predecessor in that the spaces that could not be filled from the Census data have been filled with supplementary observations from the AISA and interpolations. It was felt that these methods were too imprecise for use in Table C.9, but that they were suitable for the more approximate Table C.10. The AISA data and the Census data are not strictly comparable as the former were collected on a calendar year basis, and interpolation is always approximate. The former difficulty is mitigated by the interest in proportions, which are much more stable than quantities; the latter is offset by the interest in trends rather than precise values.

The figures in Table C.10 that were not derived from Table C.9 have the nature of their origins indicated in footnotes. (The share of "other products" is a residual in all cases.) The first part of the table contains AISA data on the production of wire rods in 1889, data which were also used in the other parts of Table C.2. (The data came from *AISA*, (1904), p. 98.) There is no information on the volume or composition of wire rod production before 1889 when, as the table shows, they were already made almost completely from steel. Wire rods were produced earlier, although in small quantities, and the switch to steel must have taken place before 1889. The relationship between the rise in the proportion of wire rods made of steel and the growth of the importance of wire rods in total production can only be surmised, but it would appear that they both occurred in the course of the 1880's.

The allocation of total rolled iron and steel production in 1869 was not given in the Census of that year, the production of steel not being identified by product. As, however, iron represented 97 per cent of the total production, the composition of iron and steel production combined could not have differed too much from that of iron alone, and most products had to be made completely or almost completely of iron. It was assumed that the small quantity of steel made in 1869 was used mainly for rails and bars; the new Bessemer steel accounted for about one-third of production and was used primarily for rails, while the remainder of production was mostly crucible steel which was produced in the form of small, simple shapes for further manufacture into high-quality products. Under this assumption, the quantity of steel made was too small to have any effect on the distribution of total production among products, and the first row of the first part of Table C.10 is the same as the first row of the third part.

It will be noted that the share of rails in the total production of rolled iron and steel in 1869 is reported differently in Tables C.7 and C.10. The differences between these two tables for years other than 1869 is well within the error to be expected from the difference between Census and calendar years, but the difference for 1869 — 10 per cent — is not. This 10 per cent discrepancy derives from the quantity data on the production of rolled iron and iron rails in Tables C.6 and C.9, the quantity of steel rails produced in 1869 being too small to be important here. It results, in all probability, from a failure of the AISA to measure accurately the production of rolled iron other than rails. They could easily have failed to report the production of small, non-rail rolling mills and consequently have overstated the relative importance of rails. (If the Ninth Census used gross tons, instead of net tons as assumed here, the two estimates of *rail* production in 1869 are quite close.) The evidence for this interpretation, however, is not strong, and the discrepancy was allowed to stand.

The same type of data used to complete the first part of Table C.10 sufficed to fill in those gaps in the second part for 1869 and 1889 that could be filled with confidence. As Table C.9 shows, the Census did not collect data for iron and steel separately after 1889, with the exception of rails and structural shapes. The AISA collected data on individual products by material used for a short time around 1890 and for the years 1904 through 1911, but with large gaps in the data for 1904 and 1909. [*AISA* (1887), p. 33; (1896), p. 72; (1904), p. 60; (1905), p. 68; (1906), p. 69; (1907), p. 75; (1909), p. 89; (1910), p. 87; (1911), II, 16.] The AISA data could be used to fill in the 1909 row of the second part of Table C.2, but not the 1899 row, if the gaps could be filled. The stability of proportions was called upon to justify some interpolation from adjacent years, and the 1909 row was filled in as shown. The AISA also supplied data from which the share of iron in the total product of the industry could be estimated for 1899, and this too was entered in the table. (The estimate appears in Table C.6 also.)

The rest of the 1899 row was filled in by interpolation. The method of interpolation rested on the interdependence of the four parts of Table C.10 and the necessity of the rows of three of them to add to 100 per cent. This interdependence served to check the interpolations themselves at the same time as it enabled them to be used to complete the last two parts of the table. The computation scheme provided the method of calculation used for these two parts for both 1899 and 1909.

The distribution of iron and steel taken together was known for 1899 (the first part of Table C.10), as was the share of iron in the total production (the last column of the second part of Table C.10). If the proportion of a product made of iron (the second part of Table C.10) was interpolated, the proportion of rolled iron used for this product (the third part of Table C.10) could be calculated by multiplying the interpolated value by the following ratio: the proportion of total production used for this product divided by the proportion of total production made of iron. This was done for all products, and an exactly analogous derivation was used to calculate the proportion of rolled steel used for each product (the fourth part of Table C.10). The rows of the calculated third and fourth parts of Table C.10 were then added together to see if they were near, but under 100 per cent. Finally, any necessary modifications in the interpolations were made. (The formulas are as follows, where P_I is the quantity of the product under discussion that was made of iron; P_S, the quantity of the product made of steel; T_I, the total amount of rolled iron; and T_S, the total amount of rolled steel.)

$$\frac{P_I}{T_I} = \frac{P_I}{P_I + P_S}\left[\left(\frac{P_I + P_S}{T_I + T_S}\right)\left(\frac{T_I + T_S}{T_I}\right)\right]$$

$$\frac{P_S}{T_S} = \left(1 - \frac{P_I}{P_I + P_S}\right)\left[\left(\frac{P_I + P_S}{T_I + T_S}\right)\left(\frac{T_I + T_S}{T_S}\right)\right]$$

This method yielded the values shown. Not all products had to be estimated, and not all of those that were estimated were completely unknown. The share of iron in rails and in structural shapes for 1899 was given in the Census. The share of iron in wire rods and nail plate did not change greatly between 1889 and 1909, and the interpolated figures were chosen to be in the same ranges. The interpolation was important and difficult only in the case of bars and rods, plates and sheets, and skelp. It was important to provide a sense of the timing of shifts in the composition of production, although the allocation of changes between two adjacent decades may not be of crucial importance. It was difficult because of the lack of information, but the constraints present reduced the chances for error considerably. If ranges were to be attached to these interpolations corresponding to some high degree of confidence in them, they would extend about five percentage points to either side of the estimates.

The results of these operations are that Table C.10 shows the changes in the composition of rolled iron and steel products from 1869 to 1909 with important gaps unfilled only for 1869 and for wire rods in 1879. These gaps are unfortunate, but not too important as the production of steel and of wire rods in these years was not large.

Table C.11: Table C.9.

Table C.12: 1839: U.S. Census, Sixth, "Compendium" (Washington, 1841), p. 358.

1849: Seventh Census, "Compendium," pp. 181-182.

1859: Eighth Census, III, clxxx, clxxxiii.

1869: Ninth Census, III, 603-607.

1879: Tenth Census, II, 746-760.

1889: Eleventh Census, VI, Part III, 400, 421.

1899: Twelfth Census, X, 33-34, 64.

1909: Thirteenth Census, X, 210, 252-253.

Table C.13: Pig iron: *AISA* (1911), I, 83.

Rails, before 1840: included in rolled bars.

Rails, 1840-1863: *AISA* (1911), I, 88.

Bars: Frank W. Taussig, "The Tariff, 1830-1860," *Quarterly Journal of Economics*, 2 (April, 1888), 379.

Table C.14: Pig iron, 1860-1870: *AISA* (1911), I, 83-84 (converted to calendar years by averaging).

Rails, 1860-1866: *AISA* (1889), p. 66 (converted to calendar years by averaging).

Rails, 1867-1870: *AISA* (1896), p. 68.

1871-1899: *AISA* (1880), p. 16; (1885), p. 14; (1887), p. 16; (1889), p. 24; (1890), p. 26; (1892), p. 26; (1894), p. 28; (1896), p. 30; (1898), p. 33; (1902), p. 21.

Table C.15: *AISA* (1896), pp. 84-88; (1899), p. 29; (1904), p. 29; (1907), p. 39; (1911), I, 39.

TABLE C.1
PIG IRON PRODUCTION BEFORE 1860
(gross tons)

	AISA	Carey	Grosvenor	Pearse	Raymond
1810	53,908	54,000		53,908¾	
1811					
1812					
1813					
1814					
1815					
1816					
1817					
1818					
1819					
1820	20,000				
1821					
1822					
1823					
1824					
1825					
1826					
1827					
1828	130,000	130,000		123,404	130,000
1829	142,000	142,000		134,954	142,000
1830	165,000	165,000		180,598	165,000
1831	191,000	191,000		191,536	191,000
1832	200,000	200,000		200,000	200,000
1833					218,000
1834					236,000
1835					254,000
1836					272,000
1837				250,000	290,000
1838					308,000
1839				235,000	326,000
1840	286,903	315,000	347,000	317,306	347,000
1841			360,000		290,000
1842	215,000	200-230,000	376,000	215,000	230,000
1843			386,000		312,000
1844			427,000	486,000	394,000
1845			486,000	502,000	486,000
1846	765,000	765,000	551,000	765,000	765,000
1847	800,000	800,000	597,674	800,000	800,000
1848	800,000	800,000	570,000	800,000	800,000
1849	650,000	650,000	542,903	650,000	650,000

TABLE C.1 – *Continued*
PIG IRON PRODUCTION BEFORE 1860
(gross tons)

	AISA	Carey	Grosvenor	Pearse	Raymond
1850	563,755		564,000	563,755	563,755
1851				513,000	413,000
1852	500,000			540,755	540,755
1853				805,000	723,214
1854	657,337			724,833	662,216
1855	700,154			728,973	700,157
1856	788,515			812,917	788,515
1857	712,640			798,157	712,640
1858	629,548			705,094	629,552
1859	750,560			840,627	750,560

TABLE C.2
PRODUCTION OF PIG IRON BY FUEL USED
(thousand gross tons)

	Bituminous Coal and Coke	Anthracite and Coke	Anthracite	Charcoal	Total
1854	49	303		306	657
1855	56	341		304	700
1856	62	396		331	789
1857	69	349		295	713
1858	52	323		255	630
1859	76	421		254	751
1860	109	464		249	821
1861	113	365		174	653
1862	117	420		167	703
1863	141	516		189	846
1864	188	611		216	1014
1865	169	428		234	832
1866	240	669		297	1206
1867	285	713		307	1305
1868	304	797		330	1431
1869	494	867		350	1711
1870	509	830		326	1665
1871	509	854		344	1707
1872	879	1223		447	2549
1873	873	1172		516	2561
1874	813	1073		515	2401
1875	846	811		367	2024
1876	884	709		276	1869
1877	948	835		284	2067
1878	1063	976		262	2301
1879	1285	1137		320	2742
1880	1741	1614		480	3835
1881	2025	1549		570	4144
1882	2177	1823		623	4623
1883	2401	821	862	510	4596
1884	2272	1196	221	409	4098
1885	2389	1050	248	357	4045
1886	·3398	1479	396	410	5683
1887	3813	1714	374	516	6417
1888	4236	1471	248	535	6490
1889	5314	1407	307	575	7604

TABLE C.2 — *Continued*
PRODUCTION OF PIG IRON BY FUEL USED
(thousand gross tons)

	Bituminous Coal and Coke	*Anthracite and Coke*	*Anthracite*	*Charcoal*	*Total*
1890	6388	1937	249	628	9203
1891	5837	1560	306	577	8280
1892	6822	1568	229	538	9157
1893	5390	1298	50	387	7125
1894	5520	795	120	222	6657
1895	7950	1214	57	225	9446
1896	7166	1035	112	310	8623
1897	8465	912	21	255	9653
1898	10274	1181	22	297	11774
1899	11736	1559	41	285	13621
1900	11728	1636	41	385	13789
1901	13782	1669	44	383	15878
1902	16316	1096	19	391	17821
1903	15592	1864	47	506	18009
1904	14931	1197	31	338	16497
1905	20965	1644	30	353	22992
1906	23313	1536	25	433	25307
1907	23972	1335	36	437	25781
1908	15332	353	2	249	15936
1909	24721	682	16	376	25795
1910	26258	629	21	397	27304
1911	23141	213	17	279	23650

TABLE C.3

PROPORTION OF PIG IRON MADE WITH DIFFERENT FUELS
(percentages)

	Bituminous Coal and Coke	Mixed Anthracite and Coke	Anthracite	Charcoal
1854	7	46		47
1855	8	49		43
1856	8	50		42
1857	10	49		41
1858	8	51		41
1859	10	56		34
1860	13	57		30
1861	17	56		27
1862	17	60		24
1863	17	61		22
1864	19	60		21
1865	20	52		28
1866	20	55		25
1867	22	55		24
1868	21	56		23
1869	29	51		20
1870	31	50		20
1871	30	50		20
1872	34	48		18
1873	34	46		20
1874	34	45		21
1875	42	40		18
1876	47	38		15
1877	46	40		14
1878	46	42		11
1879	47	41		12
1880	45	42		13
1881	49	37		14
1882	47	40		13
1883	52	18	19	11
1884	58	29	5	10
1885	59	26	6	9
1886	60	26	7	7
1887	60	27	6	8
1888	65	23	4	8
1889	70	18	4	8

TABLE C.3—Continued
PROPORTION OF PIG IRON MADE WITH DIFFERENT FUELS
(percentages)

	Bituminous Coal and Coke	Mixed Anthracite and Coke	Anthracite	Charcoal
1890	69	21	3	7
1891	70	19	4	7
1892	75	17	3	6
1893	76	18	1	5
1894	83	12	2	3
1895	84	13	1	2
1896	83	12	1	3
1897	88	9	0	3
1898	87	10	0	3
1899	86	11	0	2
1900	85	12	0	3
1901	87	11	0	2
1902	92	6	0	2
1903	87	10	0	3
1904	91	6	0	2
1905	91	7	0	2
1906	92	6	0	2
1907	93	5	0	2
1908	96	2	0	2
1909	96	3	0	1
1910	96	2	0	1
1911	98	1	0	1

TABLE C.4
STEEL PRODUCTION BY PROCESS USED
(thousand gross tons)

| | Ingots and Castings | | | | Castings | | |
| | Total | Bessemer | Open-Hearth | | Total | Open-Hearth | |
			Total	Acid		Total	Acid
1867	20	2	—				
1868	27	8	—				
1869	31	11	1				
1870	69	38	1				
1871	73	40	2				
1872	143	107	3				
1873	199	152	3				
1874	216	171	6				
1875	390	335	8				
1876	533	470	19				
1877	570	501	22				
1878	732	654	32				
1879	935	829	50				
1880	1247	1074	101				
1881	1588	1374	131				
1882	1737	1515	143				
1883	1674	1477	119				
1884	1551	1376	118				
1885	1712	1519	133				
1886	2563	2269	219				
1887	3339	2936	322				
1888	2899	2511	314				
1889	3386	2930	375				
1890	4277	3689	513				
1891	3904	3247	580				
1892	4928	4168	670				
1893	4020	3216	738				
1894	4412	3571	785				
1895	6115	4909	1137				
1896	5282	3920	1299				
1897	7157	5475	1609				
1898	8933	6609	2230	661	132	121	92
1899	10640	7586	2947	867	181	170	130

TABLE C.4—Continued
STEEL PRODUCTION BY PROCESS USED
(thousand gross tons)

	Ingots and Castings				Castings		
	Total	Bessemer	Open-Hearth		Total	Open-Hearth	
			Total	Acid		Total	Acid
1900	10188	6685	3398	853	193	177	135
1901	13474	8713	4656	1037	318	302	207
1902	14947	9138	5688	1191	391	368	255
1903	14535	8593	5830	1095	430	400	265
1904	13860	7859	5908	802	330	303	204
1905	20024	10941	8971	1156	561	527	320
1906	23398	12276	10980	1322	774	720	406
1907	23363	11668	11550	1270	803	747	380
1908	14023	6117	7837	696	346	312	157
1909	23955	9331	14494	1076	656	601	295
1910	26095	9413	16505	1212	941	863	429
1911	23676	7948	15599	913	647	571	305

TABLE C.5
PROPORTIONS OF STEEL MADE BY DIFFERENT PROCESSES
(percentages)

	Ingots and Castings			Total Castings	O–H Castings	Acid O–H Castings
	Bessemer Total	O–H Total	Acid O–H Total O–H	Total I and C	O–H I and C	Acid O–H I and C
1867	10	—				
1868	30	—				
1869	35	3				
1870	55	1				
1871	55	3				
1872	75	2				
1873	76	2				
1874	81	3				
1875	86	2				
1876	88	4				
1877	86	4				
1878	89	4				
1879	89	5				
1880	86	8				
1881	87	8				
1882	87	8				
1883	88	7				
1884	89	8				
1885	89	8				
1886	89	9				
1887	88	11				
1888	87	11				
1889	86	11				
1890	86	12				
1891	83	15				
1892	85	14				
1893	80	18				
1894	81	18				
1895	80	19				
1896	74	25				
1897	77	22				
1898	74	25	30	1	5	14
1899	71	28	29	2	6	15

TABLE C.5—Continued
PROPORTIONS OF STEEL MADE BY DIFFERENT PROCESSES
(percentages)

	Ingots and Castings			Total Castings Total	O−H Castings O−H	Acid O−H Castings Acid O−H
	Bessemer Total	O−H Total	Acid O−H Total O−H	Total O−H I and C	O−H I and C	Acid O−H I and C
1900	66	33	25	2	5	16
1901	65	35	22	2	6	20
1902	61	38	21	3	6	21
1903	56	40	19	3	7	24
1904	57	43	14	2	5	25
1905	55	45	13	3	6	28
1906	52	47	12	3	7	31
1907	50	49	11	3	6	30
1908	44	56	9	2	4	23
1900	39	60	7	3	4	27
1910	36	63	7	4	5	36
1911	34	66	6	3	4	33

O−H = Open-Hearth
I and C = Ingots and Castings

TABLE C.6
PRODUCTION OF ROLLED IRON AND STEEL
(thousand gross tons)

	Total Rolled Products	Rolled Iron	Rolled Steel	Total Rails	Iron Rails	Steel Rails
1849					22	
1850					39	
1851					45	
1852					56	
1853					78	
1854					96	
1855					124	
1856					161	
1857					145	
1858					146	
1859					175	
1860					183	
1861					169	
1862					191	
1863					246	
1864		779			299	
1865		765			318	
1866		916			385	
1867	945	928	17	413	410	2
1868	1003	980	23	452	446	6
1869	1121	1095	26	530	521	9
1870	1210	1152	58	554	523	30
1871	1353	1292	61	693	658	34
1872	1769	1650	119	893	809	84
1873	1806	1641	165	795	680	115
1874	1691	1513	178	651	522	129
1875	1749	1428	321	708	448	260
1876	1784	1347	437	785	417	368
1877	1785	1319	466	683	297	386
1878	1986	1389	597	788	288	500
1879	2589	1828	761	994	375	619

TABLE C.6 — Continued
PRODUCTION OF ROLLED IRON AND STEEL
(thousand gross tons)

	Total Rolled Products	Rolled Iron	Rolled Steel	Total Rails	Iron Rails	Steel Rails
1880	3095	2083	1012	1305	441	864
1881	3647	2361	1286	1647	436	1210
1882	3629	2227	1402	1508	203	1304
1883	3445	2097	1348	1215	58	1157
1884	2993	1748	1245	1022	23	999
1885	2082	1611	1371	977	13	964
1886	4331	2039	2292	1601	21	1579
1887	5236	2310	2925	1140	21	2119
1888	4617	2153	2464	1404	13	1391
1889	5237	2309	2928	1522	9	1513
1890	6023	2518	3505	1885	14	1871
1891	5391	2316	3075	1307	8	1299
1892	6166	2296	3870	1552	10	1541
1893	4976	1828	3148	1136	6	1130
1894	4642	1196	3446	1022	5	1017
1895	6190	1428	4762	1306	6	1300
1896	5516	1414	4102	1122	4	1117
1897	7002	1460	5542	1648	3	1645
1898	8513	1615	6898	1981	3	1978
1899	10294	2101	8193	2273	2	2271
1900	9487	1665	7822	2386	1	2385
1901	12349	2033	10316	2875	2	2873
1902	13944	2533	11411	2948	7	2941
1903	13208	2144	11064	2992	1	2992
1904	12013	1760	10253	2285	1	2284
1905	16840	2060	14780	3376	0	3376
1906	19588	2187	17402	3978	0	3978
1907	19865	2200	17665	3634	1	3633
1908	11828	1238	10590	1921	0	1921
1909	19645	1709	17935	3024	0	3024
1910	21621	1740	19880	3636	0	3636
1911	19039	1461	17579	2823	0	2823

TABLE C.7
DISTRIBUTION OF ROLLED IRON AND STEEL
(percentages)

	Rolled Iron / Total Rolled Products	Iron Rails / Total Rails	Total Rails / Total Rolled Products	Iron Rails / Rolled Iron	Steel Rails / Rolled Steel
1864				38	
1865				42	
1866				42	
1867	98	97	44	44	12
1868	98	99	45	45	26
1869	98	98	47	48	35
1870	95	94	46	45	52
1871	95	95	51	51	56
1872	93	91	50	49	71
1873	91	86	44	41	70
1874	89	80	38	34	72
1875	81	63	40	31	81
1876	75	53	44	31	84
1877	74	44	38	23	83
1878	70	38	40	21	84
1879	70	38	38	21	81
1880	67	34	42	21	85
1881	65	27	45	18	94
1882	61	13	42	9	93
1883	61	5	35	3	86
1884	58	2	34	1	80
1885	54	1	33	1	70
1886	47	1	37	1	69
1887	44	2	22	1	72
1888	47	1	30	1	56
1889	44	1	29	0	52
1890	42	1	31	1	53
1891	43	1	24	0	42
1892	38	1	25	0	40
1893	38	1	23	0	36
1894	26	0	22	0	29
1895	23	0	21	0	27
1896	26	0	20	0	27
1897	21	0	23	0	30
1898	19	0	23	0	29
1899	20	0	22	0	28

TABLE C.7–Continued
DISTRIBUTION OF ROLLED IRON AND STEEL
(percentages)

	Rolled Iron / Total Rolled Products	Iron Rails / Total Rails	Total Rails / Total Rolled Products	Iron Rails / Rolled Iron	Steel Rails / Rolled Steel
1900	18	0	25	0	31
1901	16	0	23	0	28
1902	18	0	21	0	26
1903	16	0	23	0	27
1904	15	0	19	0	22
1905	12	0	20	0	23
1906	11	0	20	0	22
1907	11	0	18	0	21
1908	10	0	16	0	18
1909	9	0	15	0	17
1910	8	0	17	0	18
1911	8	0	15	0	16

TABLE C.8
RELATIVE VOLUMES OF PRODUCTION

	(Thousand Gross Tons)				(Percentages)		
	Pig Iron Produced	Pig Iron Used for Rolled Products	Rolled Products	Ingots and Castings	Pig Used / Pig Made	Pig Used / Rolled Products	Rolled Prod. / Pig Made
1849	565	251	278	—	44	90	49
1859	988	NSR	513	11	—	—	52
1869	1833	1004	1356	45	55	74	74
1879	3376	2285	2978	1027	68	77	88
1889	8845	5847	5721	4175	66	102	65
1899	14452	10410	10399	10685	72	100	72
1909	25652	19077	19276	23523	74	99	75
1879 RM		1410	2100			67	
1879 SW		942	878			93	

NSR = Not Separately Reported.
RM = Rolling Mills.
SW = Steelworks.

TABLE C.9
(thousand gross tons)

	Rails	Bars and Rods	Plates and Sheets	Wire Rods	Skelp, etc.	Structural Shapes	Nail Plate	Other	Total
I. Total Quantity of Production of Rolled Iron and Steel Products									
1869	–	–	–	–	–	–	–	–	1369
1879	1087	874	262	–	201	88	226	184	2922
1889	1868	1572	653	–	546	276	261	830	6008
1899	2251	2493	1882	917	1195	857	98	706	10399
1909	2859	3976	3964	2295	2425	2124	69	1564	19276
II. Quantity of Rolled Products Made of Iron									
1869	475	460	138	–	2	–	205	44	1324
1879	417	722	249	–	201	87	226	142	2044
1889	13	1164	324	–	526	123	96	254	2500
1899	1	–	–	–	–	27	–	–	–
1909	0	–	–	–	–	21	–	–	–
III. Quantity of Rolled Products Made of Steel									
1869	–	–	–	–	–	–	–	–	45
1879	671	152	13	–	–	1	–	42	878
1889	1854	408	330	–	21	154	165	576	3508
1899	2250	–	–	–	–	830	–	–	–
1909	2859	–	–	–	–	2102	–	–	–
IV. Quantity of Rolled Products Made of *Bessemer* Steel									
1869	–	–	–	–	–	–	–	–	17
1879	662	112	1	–	–	0	–	18	795
1889	1854	304	145	–	17	85	165	421	2991
1899	–	–	–	–	–	264	–	–	–
1909	1644	–	–	–	–	168	–	–	–
V. Quantity of Rolled Products Made of *Open-Hearth* Steel									
1869	–	–	–	–	–	–	–	–	0
1879	8	40	11	–	–	0	–	24	83
1889	0	104	186	–	3	68	1	154	516
1899	–	–	–	–	–	566	–	–	–
1909	1215	–	–	–	–	1934	–	–	–

TABLE C.10

(percentages)

	Rails	Bars and Rods	Plates and Sheets	Wire Rods	Skelp, etc.	Structural Shapes	Nail Plate	Other	Total
I. Distribution of Total Rolled *Iron and Steel* Among Various Products									
1869	36[b]	35[b]	10[b]	—	0[b]	—	15[b]	4	100
1879	37	30	9	—	7	3	8	6	100
1889	31	26	11	7[a]	9	5	4	7	100
1899	22	24	18	9	11	8	1	7	100
1909	15	21	21	12	13	11	0	7	100
II. Proportion of Each Product Made of Iron									
1869	96[a]	96[b]	100[b]	—	100[b]	—	100[b]	—	97
1879	38	83	95	—	100	99	100	—	70
1889	1	74	50	4[a]	96	45	37	—	42
1899	0	48[b]	10[b]	0[b]	45[b]	3	33[b]	—	20[a]
1909	0	28[a]	2[a]	0[a]	18[a]	1	30[a]	—	9[a]
III. Distribution of Total Rolled *Iron* among Various Products									
1869	36	35	10	—	0	—	15	4	100
1879	20	35	12	—	10	4	11	8	100
1889	1	47	13	1[a]	21	5	4	8	100
1899	0[c]	58[c]	9[c]	0[c]	25[c]	1[c]	2[c]	5	100
1909	0[c]	65[c]	5[c]	0[c]	26[c]	1[c]	1[c]	2	100
IV. Distribution of Total Rolled *Steel* among Various Products									
1869	—	—	—	—	—	—	—	—	—
1879	76	17	1	—	0	0	0	6	100
1889	53	12	9	12[a]	1	4	5	4	100
1899	28[c]	16[c]	20[c]	11[c]	8[c]	10[c]	1[c]	6	100
1909	16[c]	17[c]	23[c]	13[c]	12[c]	12[c]	0[c]	7	100

[a] From AISA.
[b] Interpolated.
[c] Computed.
— NSR.

TABLE C.11
(percentages)

	Rails	Bars and Rods	Plates and Sheets	Wire Rods	Skelp, etc.	Struc- tural Shapes	Nail Plate	Other	Total
Proportion of Steel Product Made of Bessemer Steel									
1879	99	74	8	—	—	100	—	—	91
1889	100	75	44	—	81	55	100	—	85
1899	—					32			
1909	58					8			
Distribution of Bessemer Steel among Various Products									
1879	83	14	0	—	0	0	0	3	100
1889	62	10	5	—	1	3	0	19	100
Distribution of Open-Hearth Steel among Various Products									
1879	10	48	13	—	0	0	0	29	100
1889	0	20	36	—	1	13	0	30	100

TABLE C.12
PROPORTION OF PRODUCTION IN THE FIRST FIVE STATES
(percentages)

	First State		Second State		Third State		Fourth State		Fifth State	
Blast Furnaces										
1839	Pa.	34	Ohio	12	Ky.	10	N.Y.	10	Va.	7
1849	Pa.	51	Ohio	9	Md.	8	Tenn.	5	Ky.	4
1859	Pa.	58	Ohio	12	N.Y.	8	N.J.	5	Ky.	3
1869	Pa.	52	Ohio	15	N.Y.	11	Mo.	4	Mich.	4
1879	Pa.	51	Ohio	15	N.Y.	8	N.J.	4	Mich.	3
1889	Pa.	49	Ohio	14	Ala.	9	Ill.	8	N.Y.	3
1899	Pa.	47	Ohio	18	Ill.	10	Ala.	8	Va.	3
1909	Pa.	43	Ohio	21	Ill.	10	Ala.	7	N.Y.	7
Steelworks and Rolling Mills										
1839	Pa.	44	N.Y.	27	Tenn.	5	Md.	4	Ohio	4
1849	Pa.	66	Va.	5	Ohio	5	N.Y.	5	Tenn.	4
1859	Pa.	52	Mass.	8	Ohio	8	N.Y.	7	N.J.	6
1869[v]	Pa.	47	N.Y.	12	Ohio	10	Mass.	5	N.J.	4
1879	Pa.	49	Ohio	11	Ill.	9	N.Y.	7	Mass.	4
1889	Pa.	53	Ohio	14	Ill.	11	W. Va.	3	N.Y.	3
1899	Pa.	56	Ohio	18	Ill.	10	Ind.	3	W. Va.	2
1909	Pa.	49	Ohio	22	Ill.	10	Ind.	4	N.Y.	4

v = Value.

TABLE C.13
IMPORTS (FISCAL YEARS)
(thousand gross tons)

	Pig Iron	Iron Rails	Rolled Bars	Hammered Bars
1830	1	7		31
1831	7	15		23
1832	10	21		38
1833	9	28		36
1834	11	29		32
1835	12	28		31
1836	9	47		33
1837	14	48		31
1838	12	36		21
1839	13	60		36
1840	6	29	4	29
1841	12	23	40	30
1842	19	25	37	19
1843	4	10	6	6
1844	15	16	24	12
1845	28	22	2	18
1846	24	6	18	21
1847	28	14	26	15
1848	52	29	53	20
1849	106	62	111	11
1850	75	142	106	15
1851	67	189	65	20
1852	92	246	45	44
1853	114	299	88	18
1854	160	283	46	14
1855	99	128	117	
1856	59	155	108	
1857	52	179	87	
1858	42	76	66	
1859	73	70	95	
1860	71	122	106	
1861	74	74		
1862	22	9		
1863	31	17		
1864	102			
1865	112			

TABLE C.14
IMPORTS (CALENDAR YEARS)
(thousand gross tons)

	Pig Iron	Rails	Bars	Tin and Terne Plate	Scrap
1860	73	98			
1861	48	42			
1862	27	13			
1863	67	68			
1864	77	99			
1865	77	78			
1866	107	87			
1867	112	146			
1868	125	223			
1869	145	280			
1870	160	356			
1871	219	506	110	83	197
1872	264	474	80	86	248
1873	138	231	56	97	97
1874	55	97	24	80	36
1875	75	17	25	91	26
1876	74	0	24	90	13
1877	60	0	27	112	10
1878	67	0	30	108	6
1879	304	39	44	154	222
1880	701	259	113	158	620
1881	465	345	43	183	135
1882	540	200	71	214	147
1883	323	35	42	221	64
1884	184	3	37	216	34
1885	147	2	31	229	16
1886	362	42	29	258	97
1887	468	138	36	284	341
1888	197	63	32	298	54
1889	149	6	30	331	38
1890	135	0	24	329	56
1891	67	0	18	328	44
1892	70	0	19	268	29
1893	54	3	15	253	6
1894	16	0	9	215	2
1895	53	1	20	220	6
1896	56	8	16	119	8
1897	19	0	13	84	2
1898	25	0	19	67	2
1899	40	2	20	59	11

TABLE C.15
PRICES
(dollars per gross ton)

	No. 1 Foundry Pig at Phila.	Best Refined Rolled Bar Iron at Stores at Phila.	Bes. Steel Rails at Works in Pa.	Gray Forge Pig Iron Lake Ore, at Pitts.	Iron Rails at Mills in Pa.	No. 1 Charcoal Foundry Pig at Phila.	Hammered Bar Iron at Phila.
1840						32.75	90.00
1841						28.50	85.00
1842	25.60					28.00	83.50
1843						26.75	77.50
1844	25.75	85.62				28.25	75.00
1845	29.25	13.75				32.25	
1846	27.88	91.66				31.25	
1847	30.25	86.04			69.00	31.50	
1848	26.50	79.33			62.25	28.50	
1849	22.75	67.50			53.88	24.50	
1850	20.88	59.54			47.88		
1851	21.38	54.66			45.63		
1852	22.63	58.79			48.38		
1853	36.12	83.50			77.25		
1854	36.88	91.33			80.13		
1855	27.75	74.58			62.88		
1856	27.12	73.75			64.38		
1857	26.38	71.04			64.25		
1858	22.25	62.29			50.00		
1859	23.38	60.00			49.38		
1860	22.75	58.75			48.00		
1861	20.25	60.83			42.38		
1862	23.88	70.42			41.75		
1863	35.25	91.04			76.88		
1864	59.25	146.46			126.00		
1865	46.12	106.38			98.63		
1866	46.88	98.13			86.73		
1867	44.12	87.08	166.00		83.13		
1868	39.25	85.63	158.50		78.88		
1869	40.63	81.66	132.25		77.25		

TABLE C.15A
PRICES
(dollars per gross ton)

	No. 1 Foundry Pig at Phila.	Best Refined Rolled Bar Iron at Stores at Phila.	Bes. Steel Rails at Works in Pa.	Gray Forge Pig Iron at Lake Ore, at Pitts.	Iron Rails at Mills in Pa.	Gray Forge Pig Iron at Phila.	Steel Billets at Mills at Pitts.	Bes. Pig Iron at Pitts.
1870	33.25	78.96	106.75		72.25			
1871	35.12	78.54	102.50		70.38			
1872	48.88	97.63	112.00		85.13			
1873	42.75	86.43	120.50	35.80	76.67			
1874	30.25	67.95	94.25	27.16	58.75			
1875	25.50	60.85	68.75	23.67	47.75			
1876	22.25	52.08	59.25	21.74	41.25			
1877	18.88	45.55	45.50	20.60	35.25			
1878	17.63	44.24	42.25	18.04	33.75			
1878	21.50	51.85	48.25	22.15	41.25			
1880	28.50	60.38	67.50	27.98	49.25			
1881	25.15	58.05	61.13	22.94	47.13			
1882	25.75	61.41	48.50	23.84	45.50	22.60		
1883	22.38	50.30	37.75	19.04		19.33		
1884	19.88	44.05	30.75	17.17		17.71		
1885	18.00	40.32	28.50	15.27		15.58		
1886	18.71	43.12	34.50	16.58		16.40	31.75	18.96
1887	20.92	49.37	37.08	19.02		17.79	32.55	21.37
1888	18.88	44.99	29.83	15.99		16.21	28.78	17.38
1889	17.75	43.40	29.25	15.37		15.48	29.45	18.00
1890	18.40	45.92	31.75	15.78		15.82	30.32	18.85
1891	17.52	42.56	29.92	14.06		14.52	25.32	15.95
1892	15.75	41.81	30.00	12.81		13.54	23.63	14.37
1893	14.52	38.08	28.12	11.77		12.73	20.44	12.87
1894	12.66	29.96	24.00	9.75		10.73	16.58	11.38
1895	13.10	1.44°	24.33	10.94		11.49	18.48	12.72
1896	12.95	1.40	28.00	10.39		11.09	18.83	12.14
1897	12.10	1.31	18.75	9.03		10.48	15.08	10.13
1898	11.66	1.28	17.62	9.18		10.23	15.31	10.33
1899	19.36	2.07	28.12	16.72		16.60	31.12	19.03

TABLE C.15A
PRICES
(dollars per gross ton)

	No. 1 Foundry Pig at Phila.	Best Refined Rolled Bar Iron at Stores at Phila.	Bes. Steel Rails at Works in Pa.	Gray Forge Pig Iron at Lake Ore, at Pitts.	Iron Rails at Mills in Pa.	Gray Forge Pig Iron at Phila.	Steel Billets at Mills at Pitts.	Bes. Pig Iron at Pitts.
1900	19.98	1.96°	32.29	16.90		16.49	25.06	19.49
1901	15.87	1.84	27.33	14.20		14.08	24.13	15.93
1902	22.19	2.13	28.00	19.49		19.20	30.57	20.67
1903	19.92	2.00	28.00	17.52		17.13	27.91	18.98
1904	15.57	1.72	28.00	12.89		13.67	22.10	13.76
1905	17.88	1.92	28.00	15.62		15.58	24.03	16.36
1906	20.98	1.98	28.00	18.19		17.79	27.45	19.54
1907	23.89	2.11	28.00	21.52		21.06	29.25	22.84
1908	17.70	1.70	28.00	15.23		15.72	26.31	17.07
1909	17.81	1.76	28.00	15.55		16.13	24.62	17.41
1910	17.36	1.85	28.00	15.24		15.72	25.38	17.19
1911	15.71	1.64	28.00	13.97		14.43	21.46	15.71

° Dollars per hundred pounds.

Selected Bibliography

Periodicals

The following abbreviations were used to represent frequently cited periodicals in the text:

AIME   *Transactions of the American Institute of Mining Engineers.* New York: 1871-.

AISA   American Iron and Steel Association, *Statistics of the American and Foreign Iron Trades.* Philadelphia: 1867, 1871-1911.

Bulletin   *Bulletin of the American Iron and Steel Association.* Philadelphia: 1866-1912.

ARJ   *American Railroad Journal.* New York: 1832-.

E&MJ   *Engineering and Mining Journal.* New York: 1866-.

Hunt's   *Hunt's Merchants' Magazine.* New York: 1839-1870.

Iron Age   *The Iron Age.* New York: 1873-.

JFI   *Journal of the Franklin Institute.* Philadelphia: 1826-.

Other Sources

Alexander, J. H., *Report on the Manufacture of Iron.* Baltimore, 1840.

American Institute of Mining Engineers, *Memorial of Alexander Lyman Holley.* New York, 1884.

American Iron Association, *Bulletin.* Philadelphia, 1857-1858.

American Iron and Steel Association, *Directory to the Iron and Steel Works of the United States*. Philadelphia, 1876-1900.

———, *The Duty on Steel Rails*. Philadelphia, 1880.

———, *History of the Manufacture of Armor Plate for the United States Navy*. Philadelphia, 1899.

Armes, Ethel, *The Story of Coal and Iron in Alabama*. Birmingham, Alabama: published under the auspices of the Chamber of Commerce, 1910.

Ashton, Thomas S., *Iron and Steel in the Industrial Revolution*. Manchester: At the University Press, 1924.

Belcher, Wallace E., "Industrial Pooling Agreements," *Quarterly Journal of Economics*, 19 (November, 1904), 111-123.

Bell, I. Lowthian, *Notes of a Visit to Coal and Iron Mines and Ironworks in the United States*. Second edition; Newcastle-on-Tyne, 1875.

Berglund, Abraham, and Philip G. Wright, *The Tariff on Iron and Steel*. Washington: The Brookings Institution, 1929.

Berry, Thomas Senior, *Western Prices Before 1861*. Cambridge: Harvard University Press, 1943.

Bessemer, Sir Henry, *An Autobiography*. London, 1905.

Bining, Arthur Cecil, *British Regulation of the Colonial Iron Industry*. Philadelphia: University of Pennsylvania Press, 1933.

———, *The Pennsylvania Iron Manufacture in the Eighteenth Century*. Harrisburg: Pennsylvania Historical Commission, 1938.

Blake, Nelson Manfred, *Water for the Cities*. Syracuse, New York: Syracuse University Press, 1956.

Boucher, John Newton. *William Kelly: A True History of the So-Called Bessemer Process*. Greenburg, Pennsylvania: The Author, 1924.

Boyer, Charles S., *Early Forges and Furnaces in New Jersey*. Philadelphia: University of Pennsylvania Press, 1931.

Bridge, James Howard, *The Inside History of the Carnegie Steel Company.* New York: Aldine Book Co., 1903.

Brody, David, *Steelworkers in America, the Nonunion Era.* Cambridge: Harvard University Press, 1960.

Burgess, George H., and Miles C. Kennedy, *Centennial History of the Pennsylvania Railroad.* Philadelphia: The Pennsylvania Railroad Co., 1949.

Burn, D. L., *The Economic History of Steelmaking,* 1867-1939. Cambridge, England: At the University Press, 1940.

———, "The Genesis of American Engineering Competition, 1850-1870," *Economic History,* supplement to the *Economic Journal,* 2 (January, 1931), 292-311.

Butler, Joseph G., Jr., *Fifty Years of Iron and Steel.* Fourth edition; Cleveland: The Penton Press Co., 1923.

Campbell, Harry Huse, *The Manufacture and Properties of Iron and Steel.* Second edition; New York: The Engineering and Mining Journal, 1903.

Cappon, Lester J., "Trend of the Southern Iron Industry under the Plantation System," *Journal of Economic and Business History,* 2 (February, 1930), 353-381.

Carey, Henry C., *Harmony of Interests,* printed in *Miscellaneous Works,* Philadelphia, 1872.

———, "Review of the Report of D. A. Wells," reprinted in *Miscellaneous Works.*

Carnegie, Andrew, *Autobiography of Andrew Carnegie.* Boston: Houghton Mifflin, 1920.

———, The Papers of Andrew Carnegie, Manuscript Division, Library of Congress.

Carnegie Brothers and Company, *The Edgar Thomson Steel Works and Blast Furnaces.* Pittsburgh, 1890.

Carr, J. C., and W. Taplin, *History of the British Steel Industry.* Cambridge: Harvard University Press, 1962.

Casson, Herbert N., *The Romance of Steel.* New York: A. S. Barnes and Co., 1907.

Caves, Richard E., *Air Transport and its Regulators*. Cambridge: Harvard University Press, 1962.

Chandler, Alfred C. Jr., *Strategy and Structure: Chapters in the History of the Industrial Enterprise*. Cambridge: M.I.T. Press, 1962.

Childs, C. G., *The Coal and Iron Trade, Embracing Statistics of Pennsylvania; a Series of Articles Published in the Philadelphia Commercial List, in 1847*. Philadelphia, 1847.

Clapham, J. H., *An Economic History of Britain*. Cambridge, England: At the University Press, 1930.

Clark, Victor S., *History of Manufactures in the United States*. 3 vols.; New York: McGraw-Hill, 1929.

Colburn, Zeran, and Alexander L. Holley, *The Permanent Way and Coal-Burning Locomotive Boilers of European Railways*. New York, 1858.

Convention of Iron Masters, *Documents Relating to the Manufacture of Iron, Published on Behalf of the Convention of Iron Masters which Met in Philadelphia on the 20th of December, 1849*. Philadelphia, 1850.

Convention of Iron Workers, *Proceedings of a Convention of Iron Workers Held at Albany, N.Y., on the 12th day of December, 1849*. Albany, 1849.

Cooper, Theodore, "The Use of Steel for Bridges," *Transactions of the American Society of Civil Engineers*, 8 (October, 1879), 263-277.

Coxe, Tench, "Digest of Manufactures, communicated to the Senate on the 5th of January, 1814," *American State Papers, Finance, II*, 666-812.

Daddow, Samuel Harries, and Benjamin Bannan, *Coal, Iron, and Oil, or the Practical American Miner*. Pottsville, Pennsylvania, 1866.

Davis, A. J., *History of Clarion County, Pa*. Syracuse, 1887.

Dictionary of American Biography. New York: Charles Scribner's Sons, 1928-1936.

Durfee, W. F., "The Manufacture of Steel," *Popular Science Monthly, 39* (October, 1891), 729-749, and *40* (November, 1891), 15-40.

Eavenson, Howard N., "The Early History of the Pittsburgh Coal Bed," *Western Pennsylvania Historical Magazine, 22* (September, 1939), 165-176.

Fisher, Douglas Alan, *The Epic of Steel.* New York: Harper and Row, 1963.

Fitch, John A., *The Steel Workers.* New York: Russell Sage Foundation, 1911.

Fogel, Robert W., "Railroads and American Economic Growth: Essays in Econometric History," unpublished doctoral dissertation, Johns Hopkins University, 1963.

French, B. F., *Rise and Progress of the Iron Trade of the United States from 1621 to 1857.* New York, 1858.

Frickey, Edwin, *Production in the United States, 1860-1914.* Cambridge: Harvard University Press, 1947.

Friends of Domestic Industry, *Report of the Committee on the Product and Manufacture of Iron and Steel of the General Convention of the Friends of Domestic Industry Assembled at New York, October 26, 1831.* Baltimore, 1832.

Fritz, John, *The Autobiography of John Fritz.* New York: John Wiley and Sons, Inc., 1912.

Fry, John E., Letter to Sir Henry Bessemer, April 20, 1896, printed in *Scientific American Supplement, 41* (May 30, 1896), 17022.

Gates, Paul W., *The Farmer's Age: Agriculture, 1815-1860.* New York: Holt, Rinehart and Winston, 1960.

Giedion, Sigfried, *Space, Time and Architecture.* Fourth edition; Cambridge: Harvard University Press, 1962.

Gloag, John, and Derek Bridgwater, *A History of Cast Iron in Architecture.* London: Allen and Unwin, 1943.

Graham, H. W., *One Hundred Years.* Pittsburgh: Jones and Laughlin Steel Corp., 1953.

Gregg, Robert, *Origin and Development of the Tennessee Coal, Iron and Railroad Company*. New York: The Newcomen Society, 1948.

Grosse, R. N., "Determinants of the Size of Iron and Steel Firms in the United States, 1820-1880," unpublished doctoral dissertation, Harvard University, 1948.

Grosvenor, W. M., *Does Protection Protect?* New York, 1871.

Habakkuk, H. J., *American and British Technology in the Nineteenth Century*. Cambridge, England: At the University Press, 1962.

Hendrick, Burton J., *The Life of Andrew Carnegie*. Garden City, New York: Doubleday, Doran and Co., 1932.

Hewitt, Abram S., "The Production of Iron and Steel in Its Economic and Social Relations," *Reports of the United States Commissioners to the Paris Exposition, 1867*, Vol. II. Washington, 1870.

Holley, Alexander L., *The Bessemer Process and Works in the United States*. From the "Troy Daily Times," July 27, 1868. New York, 1868.

————, Reports to his Clients, 1874-1877.

Hull, George H., *Reasons for the Establishment of a National Pig Iron Storage Association*. Louisville, Kentucky, 1888.

Hunter, Louis C., "Factors in the Early Pittsburgh Iron Industry," in *Facts and Factors in Economic History: Articles by former students of Edwin Francis Gay*. Cambridge: Harvard University Press, 1932.

————, "Financial Problems of the Early Pittsburgh Iron Manufacturers," *Journal of Economic and Business History*, 2 (May, 1930), 520-544.

————, "The Heavy Industries Before 1860," in Harold F. Williamson (editor), *The Growth of the American Economy*. New York: Prentice-Hall, 1944.

————, "The Influence of the Market upon Technique in the Iron Industry in Western Pennsylvania Up to 1860," *Journal of Economic and Business History, 1* (February, 1929), 241-281.

————, "A Study of the Iron Industry of Pittsburgh Before 1860," unpublished doctoral dissertation, Harvard University, 1928.

Iron and Steel Institute (Great Britain), *The Iron and Steel Institute in America in 1890*. London, 1891.

Isard, Walter, "Some Locational Factors in the Iron and Steel Industry since the Early Nineteenth Century," *Journal of Political Economy*, 56 (June, 1948), 203-217.

Jeans, J. Stephen (editor), *American Industrial Conditions and Competition*. London, 1902.

Jenks, Leland Hamilton, *The Migration of British Capital to 1875*. New York: Alfred A. Knopf, 1927.

Johnson, Walter R., *Notes on the Use of Anthracite*. Boston, 1841.

Kirkland, Edward C., *Industry Comes of Age: Business, Labor, and Public Policy, 1860-1897*. New York: Holt, Rinehart and Winston, 1961.

Larsen, B. M., *A New Look at the Nature of the Open-Hearth Process*. New York: American Institute of Mining, Metallurgical and Petroleum Engineers, 1956.

Lesley, J. P., *The Iron Manufacturer's Guide to the Furnaces, Forges and Rolling Mills of the United States*. New York, 1859.

Masters, Judge Joseph, "Brief History of the early Iron and Steel Industry of the Wood, Morrell and Company and the Cambria Iron Company at Johnstown, Pa." Typescript, 1914 (Collection of E. E. Morison).

Moldenke, Richard George Gottlob, *Charcoal Iron*. Lime Rock, Connecticut: Salisbury Iron Corp., 1920.

National Association of Iron Manufacturers, *Statistical Report for 1872*. Philadelphia, 1873.

Navin, Thomas R., and Marian V. Sears, "The Rise of a Market for Industrial Securities, 1887-1902," *Business History Review*, 29 (1955), 105-138.

Nelson, Ralph L., *Merger Movements in American Industry, 1895-1956*. Princeton: Princeton University Press (NBER), 1959.

Nettels, Curtis P., *The Emergence of a National Economy*. New York: Holt, Rinehart and Winston, 1962.

Nevins, Allan, *Abram S. Hewitt*. New York: Harper and Brothers, 1935.

Newton, Joseph, *An Introduction to Metallurgy*. Second edition; New York: John Wiley and Sons, 1947.

Noble, Henry Jeffers, *History of the Cast Iron Pressure Pipe Industry in the United States of America*. New York: The Newcomen Society, 1940.

The Otis Steel Company — Pioneer, Cleveland, Ohio. Cambridge: privately printed, 1929.

Overman, Frederick, *The Manufacture of Iron*. Third edition; Philadelphia, 1854.

Pearse, John B., *A Concise History of the Iron Manufacture of the American Colonies*. Philadelphia, 1876.

Peters, Richard, Jr., *Two Centuries of Iron Smelting in Pennsylvania*. Philadelphia: Pulaski Iron Co., 1921.

Popplewell, Frank, *Some Modern Conditions and Recent Developments in Iron and Steel Production in America*. Manchester: At the University Press, 1906.

Raymond, R. W., Appendix to A. S. Hewitt, *A Century of Mining and Metallurgy in the United States*. Philadelphia, 1876.

Redlich, Fritz, *History of American Business Leaders*. Ann Arbor, Michigan: Edwards Brothers, 1940.

Reeves, Samuel J., Letter to Wm. M. Meredith, November 21, 1849, pp. 653-656 of U.S. Treasury, *Report on the Finances, 1849*.

Rogin, Leo, *The Introduction of Farm Machinery in its Relation to the Productivity of Labor in the Agriculture of the United States During the Nineteenth Century*. Berkeley: University of California Press, 1931.

Schubert, H. R., *History of the British Iron and Steel Industry from c. 450 B.C. to A.D. 1775*. London: Routledge and Kegan Paul, 1957.

Scrivenor, Harry, *History of the Iron Trade from the Earliest Records to the Present Period.* London, 1854.

Sellew, William H., *Steel Rails.* New York: D. Van Nostrand Co., Inc., 1913.

Smith, D. P., L. W. Eastwood, D. J. Carney, and C. E. Sims, *Gases in Metals.* Cleveland: American Society for Metals, 1953.

Smolensky, Eugene, "Comment" (on Temin, "The Composition of Iron and Steel Products 1869-1909"), *Journal of Economic History, 23* (December, 1963), 472-476.

Strassmann, W. Paul, *Risk and Technological Innovation.* Ithaca: Cornell University Press, 1959.

Swank, James M., *History of the Manufacture of Iron in All Ages.* Second edition; Philadelphia, 1892.

Swineford, A. P., *History and Review of the Copper, Iron, Silver, Slate and other Material Interests of the South Shore of Lake Superior.* Marquette, Michigan, 1876.

Taussig, F. W., *Some Aspects of the Tariff Question.* Third edition; Cambridge: Harvard University Press, 1931.

———, *The Tariff History of the United States.* Fourth edition; New York, 1898.

———, "The Tariff, 1830-1860," *Quarterly Journal of Economics, 2* (April, 1888), 314-346, 379-384.

Taylor, R. C., *Statistics of Coal.* Second edition; Philadelphia, 1855.

Temin, Peter, "The Composition of Iron and Steel Products, 1869-1909," *Journal of Economic History, 23* (December, 1963), 447-471.

Thompson, Lillian Gilchrist, *Sidney Gilchrist Thomas: An Invention and its Consequences.* London: Faber and Faber, 1940.

Thurston, George H. *Pittsburgh as It Is.* Pittsburgh, 1857.

Trenton Iron Company, *Annual Report of the Secretary* (Abram S. Hewitt), 1858-1861. New York, 1858-1861.

————, *Documents Relating to the Trenton Iron Company.* New York, 1854.

U.S. Bureau of the Census, *Historical Statistics of the United States, Colonial Times to 1957.* Washington, 1960.

U.S. Census, Sixth, "Compendium" (Washington, 1841).

————, Seventh, "Compendium" (Washington, 1854).

————, Eighth, Vol. III, "Manufactures of the United States in 1860" (Washington, 1865).

————, Ninth, Vol. III, "Wealth and Industry" (Washington, 1872).

————, Tenth, Vol. II, "Manufactures" (Washington, 1883).

————, Eleventh, Vol. VI, "Manufacturing Industries," Part III, "Selected Industries" (Washington, 1895).

————, Twelfth, Vol. X, "Special Report on Selected Industries" (Washington, 1902).

————, Thirteenth, Vol. X, "Reports for Principal Industries" (Washington, 1913).

U.S. Census of Manufactures, 1905, Vol. IV, "Special Reports on Selected Industries" (Washington, 1908).

————, 1914, "Abstract" (Washington, 1917).

U.S. Commissioner of Corporations, *Report on the Steel Industry.* Washington, 1913.

U.S. Commissioner of Labor, 6th Annual Report, 1890, *Cost of Production: Iron, Steel, Coal, etc.* Washington, 1891.

U.S. Congress, House, Executive Document No. 308, 22d Cong., 1st Sess., "Documents Relative to the Manufactures in the United States," collected by the Secretary of the Treasury, Louis McLane (1833).

————, ————, Executive Document No. 21, 25th Cong., 3d Sess., "Report on the Steam Engines in the United States" (1838).

————, ————, Executive Document No. 2, 36th Cong., 2d Sess. (1860).

——, ——, Executive Document No. 27, 41st Cong., 2d Sess., "Report of the Special Commissioner of the Revenue" (David A. Wells)(1869).

——, ——, Miscellaneous Document No. 176, 51st Cong., 1st Sess. (1890).

——, ——, Miscellaneous Document No. 43, 53rd Cong., 1st Sess. (1893).

——, ——, Document No. 338, 54th Cong., 2d Sess. (1896).

——, ——, Document No. 1505, 60th Cong., 2d Sess., (1908-09).

——, Senate, Executive Document No. 444, 29th Cong., 1st Sess., "Report of the Secretary of the Treasury, July 23, 1846" (1846).

——, ——, Report No. 2332, 50th Cong., 1st Sess., (1888).

U.S. District Court (New Jersey), *United States vs. U.S. Steel et al.* (1914).

U.S. Navy Department, *Circular concerning Armor Plates and Appurtenances required under the Advertisement of the Honorable Secretary of the Navy, Dated April 15, 1896, to the Steel Manufacturers of the United States.* Washington, 1896.

U.S. Patent Office, *Decisions of the Commissioners of Patents, 1870, 1871.* Washington, 1871, 1872.

U.S. Tariff Commission, *Report.* Washington, 1882.

Wilson, John, *Special Report on the New York Industrial Exhibition.* British Parliamentary Reports, 1854, Vol. 36.

Winslow, John F., John A. Griswold, and Dan'l J. Morrell, Trustees, and Z. S. Durfee, Agent, *The Pneumatic or Bessemer Process of Making Iron and Steel.* Philadelphia, 1868.

Winslow, Griswold *et al.*, Agreement to modify the license granted by Winslow and Griswold to the Pennsylvania Steel Company in accordance with the transfer of ownership of patents in return for $1 paid to Morrell, April 14, 1866. (Collection of E. E. Morison).

———, Letter from Blatchford (attorney) to Winslow, January 22, 1867 (Collection of E. E. Morison).

———, *License for Kelly, Bessemer and Mushet Patents* (n.p., n.d.) (Collection of E. E. Morison).

Wright, Carroll D., *The History and Growth of the United States Census*. Washington, 1900.

Index